DISCARD
OTTERBEIN UNIVERSITY
COURTRIGHT MEMORIAL LIBRARY

Courtright Memorial Library
Otterbein College
Westerville, Ohio 43081

Contributions to Statistics

V. Fedorov/W.G. Müller/I.N. Vuchkov (Eds.)
Model-Oriented Data Analysis,
XII/248 pages, 1992

J. Antoch (Ed.)
Computational Aspects of Model Choice,
VII/285 pages, 1993

W.G. Müller/H.P. Wynn/A.A. Zhigljavsky (Eds.)
Model-Oriented Data Analysis,
XIII/287 pages, 1993

P. Mandl/M. Hušková (Eds.)
Asymptotic Statistics,
X/474 pages, 1994

P. Dirschedl/R. Ostermann (Eds.)
Computational Statistics,
VII/553 pages, 1994

C.P. Kitsos/W.G. Müller (Eds.)
MODA 4 – Advances in Model-Oriented Data Analysis,
XIV/297 pages, 1995

H. Schmidli
Reduced Rank Regression,
X/179 pages, 1995

W. Härdle/M.G. Schimek (Eds.)
Statistical Theory and Computational Aspects of Smoothing,
VIII/265 pages, 1996

S. Klinke
Data Structures for Computational Statistics,
VIII/284 pages, 1997

A.C. Atkinson/L. Pronzato/H.P. Wynn (Eds.)
MODA 5 – Advances in Model-Oriented Data
Analysis and Experimental Design
XIV/300 pages, 1998

M. Moryson
Testing for Random Walk Coefficients in Regression and State Space Models
XV/317 pages, 1998

S. Biffignandi (Ed.)
Micro- and Macrodata of Firms
XII/776 pages, 1999

W. Härdle/H. Liang/J. Gao
Partially Linear Models
X/203 pages, 2000

Werner G. Müller

Collecting Spatial Data

Optimum Design of Experiments for Random Fields

2nd Revised Edition

With 32 Figures and 6 Tables

Physica-Verlag
A Springer-Verlag Company

Series Editors
Werner A. Müller
Martina Bihn

Author
Dr. Werner G. Müller
University of Economics
and Business Administration Vienna
Department of Statistics
Augasse 2–6
A-1090 Vienna
Austria
E-mail: werner.mueller@wu-wien.ac.at

ISSN 1431-1968
ISBN 3-7908-1333-8 Physica-Verlag Heidelberg New York

Cataloging-in-Publication Data applied for
Die Deutsche Bibliothek – CIP-Einheitsaufnahme
Müller, Werner G.: Collecting spatial data: optimum design of experiments for random fields; with 6 tables / Werner G. Müller. – 2., rev. ed. – Heidelberg; New York: Physica-Verl., 2001
   (Contributions to statistics)
   ISBN 3-7908-1333-8

This work is subject to copyright. All rights are reserved, whether the whole or part of the material is concerned, specifically the rights of translation, reprinting, reuse of illustrations, recitation, broadcasting, reproduction on microfilm or in any other way, and storage in data banks. Duplication of this publication or parts thereof is permitted only under the provisions of the German Copyright Law of September 9, 1965, in its current version, and permission for use must always be obtained from Physica-Verlag. Violations are liable for prosecution under the German Copyright Law.

Physica-Verlag Heidelberg New York
a member of BertelsmannSpringer Science+Business Media GmbH

© Physica-Verlag Heidelberg 2001
Printed in Germany

The use of general descriptive names, registered names, trademarks, etc. in this publication does not imply, even in the absence of a specific statement, that such names are exempt from the relevant protective laws and regulations and therefore free for general use.

Softcover design: Erich Kirchner, Heidelberg

SPIN 10779944   88/2202-5 4 3 2 1 0 – Printed on acid-free paper

To Brigitte, David and Simon

Preface to the Second Edition

I was very pleased when Dr. Bihn proposed me to publish a second edition of this monograph. Not only did this allow me to correct quite a number of misprints (a few disturbing amongst them), but also to take account of the rapid current development in the area of spatial statistics (and within it the one of spatial design) by including and addressing many new relevant references (and updating older ones).

New to this addition is also the inclusion of exercises at the end of each chapter (with suggested solutions to be published on a web site). This allows a deeper practical involvement of the readers with the material and also a flexible way of extending it as new results become available. I am grateful to Dr. C. Milota from the Niederösterreichischen Landesregierung for making a corresponding data set available to me, and to G. Grafeneder for providing his initial geostatistical analysis of these data.

I would like to offer my thanks to E. Glatzer, M. Holtmann, I. Molchanov, and A. Pázman for pointing out inaccuracies that would otherwise have been left undetected. My special gratitude goes to D. Uciński for his very thorough reading of the first edition and his long list of suggested improvements.

Gramatneusiedl, July 26, 2000 *Werner G. Müller*

Preface to the First Edition

The aim of this monograph is to provide an overview over classical as well as recently developed methods for efficient collection of spatial data. In the past 10 years, starting with my involvement with the International Institute for Applied Systems Analysis (IIASA), I have devoted much of my research to this subject. This monograph is a compilation of the results of this work with additional material on kriging, variogram estimation and design techniques. Some parts of the book are adapted from papers coauthored with V.V. Fedorov, A. Pázman and D.L. Zimmerman. I am indebted to them for the fruitful cooperation and many helpful advices on the preparation of this text.

The theoretical elaborations are accompanied by an applied example, the redesign of the Upper-Austrian SO_2 monitoring network. I am most grateful to Dr. E. Danninger from the Oberösterreichischen Landesregierung for making the corresponding data available to me.

Additionally I would like to acknowledge the positive impact of my past and present working environments at the Department of Statistics of the University of Vienna, the Institute for Advanced Studies Vienna, the Department of Statistics and Actuarial Sciences of the University of Iowa and currently the Department of Statistics of the University of Economics Vienna. Many thanks to all the colleagues for their support and friendship.

My deep gratitude goes to E. Glatzer, W. Katzenbeisser, J. Ledolter, and K. Pötzelberger for attentive readings of and for detecting a number of inaccuracies in previous versions of this text. I am especially grateful to P. Hackl for his committed efforts in this respect that led to a considerable improvement of the manuscript.

Gramatneusiedl, June 3, 1998 *Werner G. Müller*

Contents

1 **Introduction** 1
   References 8
2 **Fundamentals of Spatial Statistics** 9
   2.1 Estimation of Spatial Trend 11
   2.2 Universal Kriging 12
   2.3 Local Regression 15
   2.4 Variogram Fitting 21
   2.5 Example 24
   2.6 Exercises 28
       References 32
3 **Fundamentals of Experimental Design** 37
   3.1 Information Matrices 39
   3.2 Design Criteria 44
   3.3 Numerical Algorithms 51
   3.4 Further Design Topics Useful in the Spatial Setting . . . 53
   3.5 Example 57
   3.6 Exercises 60
       References 63
4 **Exploratory Designs** 69
   4.1 Deterministic and Random Sampling 71
   4.2 Space Filling Designs 74
   4.3 Designs for Local Regression 77
   4.4 Model Discriminating Designs 80
   4.5 Example 82
   4.6 Exercises 84
       References 87

5 Designs for Spatial Trend Estimation — 91
- 5.1 Approximate Information Matrices 92
- 5.2 Replication-free Designs 95
- 5.3 Designs for Correlated Fields 100
- 5.4 Designs for Spatial Prediction 113
- 5.5 Example 115
- 5.6 Exercises 119
- References 121

6 Multipurpose Designs Including Designs for Variogram Fitting — 125
- 6.1 Designs for Variogram Estimation 127
- 6.2 Augmenting Designs 130
- 6.3 Alternative Methods which Ignore Correlations 132
- 6.4 Combining Different Purpose Designs 137
- 6.5 Example 140
- 6.6 Exercises 143
- References 147

Appendix — 151
- A.1 Data Sets 151
- A.2 Proofs for Chapter 2 153
- A.3 Proofs for Chapter 3 154
- A.4 Proofs for Chapter 4 156
- A.5 Proofs for Chapter 5 159
- A.6 Proofs for Chapter 6 172
- A.7 D2PT Description 174
- References 181

List of Figures — 183

Author Index — 185

Subject Index — 191

List of Symbols

A support of replication-free design

c covariance function

C error covariance matrix

d prediction variance

D covariance matrix of parameters

$\mathcal{D}$ discrepancy measure

$\mathcal{E}$ entropy

f arbitrary functions

F cumulative distribution function

$\mathcal{F}$ induced design space

h distance between two spatial locations

H set of distances

$\mathcal{H}$ distance matrix

i, j, k general indices

$\mathcal{I}$ indicator function

J measure of information

K observation covariance matrix

l general vector

L general matrix

$\mathcal{L}$ lag space

m dimension of parameter vector β

M information matrix

$\mathcal{M}$ set of information matrices

n number of support points

N number of observations

$\mathcal{N}$ number of candidate support points

p order of polynomial or design weights

$p(\cdot)$ probability density

$q(\cdot), r(\cdot)$ general functions

r remainder term

s index in sequential procedures

S supporting set

$\mathcal{S}$ generalized sum of squares

t temporal index

T number of time points

U c.d.f. of a uniform distribution

$\mathcal{U}$ mapped design space

w weights

V, W weighting matrices

List of Symbols

- x spatial coordinates
- X regressor matrix
- $\mathcal{X}$ design space
- y observed data
- $\mathcal{Z}$ region of interest
- α number between 0 and 1
- β parameters of spatial trend
- γ variogram
- Γ variogram matrix
- δ Kronecker function
- ε stochastic fluctuation (errors)
- ϵ small number
- ζ jump function
- η response surface
- θ parameters of the covariance function (variogram)
- κ fixed number or function
- λ eigenvalues or selectable parameters
- ξ exact design
- $\xi(\cdot)$ design measure
- π probability density
- ρ selectable parameter
- σ^2 error variance
- Σ covariance matrix of variogram estimates
- τ parameters after reparameterization
- ϕ sensitivity function
- Φ design criterion
- φ eigenfunctions or selectable functions
- ω restricting density
- Ω parameter space for β
- ∇ gradient matrix

Sub- and superscripts

- $\dot{}$ linearization
- $\bar{}$ average or discrete set
- $\tilde{},\hat{}$ estimators or estimates
- $+$ supplementary term
- $-$ generalized inverse
- $*$ optimum value
- $\#$ alternative or 'close' to optimum value
- $'$ alternative value
- T transposed
- 0 initial guesses or predetermined values
- min,max minimum, maximum value

1
Introduction

" Und Ihr könnt messen, richtig messen? "
" Ja. Ich habe mich sowohl an Horizontal- als auch an Höhenmessungen beteiligt, obgleich ich nicht behaupten will, daß ich mich als ausgelernten Feldmesser betrachte. "
May (1893)

Spatial data occur in many fields such as agriculture, geology, environmental sciences, and economics (see Isaaks and Srivastava (1989) or Haining (1990)). They have been recorded and analyzed probably as early as men started to make maps, however the origins of their statistical analysis as we understand it today must probably be attributed to the work of Matheron (1963). Spatial data has the distinctive characteristic that, attached to every observation, we have a set of coordinates that identifies the (geographical) position of a respective data collection site.

The set of locations of those data collection sites (the so-called design) influences decisively the quality of the results of the statistical analysis. Usually in choosing the design the aim is to ensure continuous monitoring of a data generating process or to allow for point prediction of present or future states of nature. Cox et al. (1997) have recently listed current and future issues in this research area, many of which will also be addressed on the following pages.

Sampling theory (see Bellhouse (1988)) and *optimum experimental design theory* (see Atkinson and Fedorov (1989)) are two large branches in theoretical statistics that have developed separately, though with considerable theoretical overlap (see e.g. Fienberg and Tanur (1987)), both of them providing methods for efficient site positioning. Whereas *sampling theory* is a basically model-free methodology essentially oriented towards restoring unobserved data, in *optimum design theory* the aim is to estimate the structure of the data generating process, e.g. the parameters of an assumed (regression) model or functions of these parameters.

In this monograph emphasis is on the latter branch but divergences and parallels between the two branches are pointed out whenever necessary. Principles from *optimum design theory* will be adhered to in adapting and developing methods specific for the solution of the spatial design problem. A complementary review of the model-free approaches to spatial design can be found in Arbia (1989). Merging the terminologies from the different fields is unavoidable, which will be achieved by putting the material into the random field perspective (cf. Christakos (1992)).

Fundamentals of Spatial Statistics

In Chapter 2 a short overview on the predominant geostatistical techniques is given from a classical statistics standpoint (cf. Cressie (1993)). Estimation techniques for parameters of measures of both first (mean, trend) and second order characteristics (variogram, spatial covariance function) are critically reviewed. Universal kriging is presented as a special form of generalized least squares estimation. A well-known non-parametric estimation method, the local regression technique, is shown to be a useful complement to kriging. Some of its properties, that are relevant in the spatial setting, are derived. Kriging requires parametric variogram estimation. It is indicated that the most frequently applied (weighted least squares) estimator is generally inconsistent and remedies for this problem are suggested.

In this monograph mainly two-dimensional spatial representations are chosen as the framework for most of the methods presented. However, straightforward generalizations to higher dimensions are often possible. Sometimes one-dimensional examples are employed for illustrative purposes.

Fundamentals of Experimental Design

Chapter 3 contains a description of the basic principles of *optimum experimental design theory*. Here, optimality refers to a prespecified criterion, such as the minimization of the variances of the estimates. Emphasis is again on aspects that are particularly relevant for spatial problems, such as localizing extremal values for which a reparameterization approach is given. Designs for random coefficient models, that are particularly useful in the presence of spatio-temporal relationships, are compared with some more traditional approaches. The techniques presented in this chapter are almost exclusively based upon the assumption of uncorrelated observations (see e.g. Pukelsheim (1993)) and the possibility of instantaneous replications. This however, is rarely the case in practice and extensions of the methods that cope with these limitations are tackled in the subsequent chapters.

Exploratory Designs

Frequently, in applied studies a priori knowledge of the form of the response (or regression) function is not available. Hence in an exploratory phase designs that do not require rigid models (as in Chapters 3, 5 and 6) are needed. Conclusions that are drawn from experiments with such designs may be less powerful, but, on the other hand, they are more robust and allow for model building and model validation. Several widely known basic techniques such as regular grids, random and stratified sampling, central composite designs are reviewed in Chapter 4 with particular emphasis on their properties in a spatial framework. Increasingly popular for higher dimensional sampling problems become the so-called space filling designs that aim to spread out intersite distances. A new construction rule for such designs is given and compared with more traditional approaches (like maximin or minimax distance designs). Another method that has value as an exploratory tool is local regression, for which a design criterion (the minimization of the average variance of prediction) is formulated. An optimum design strategy based upon it is given accordingly.

Designs for Spatial Trend Estimation

The two most distinguishing features of spatial designs are that observations may not be replicated instantaneously and that they are usually correlated across locations. For the former problem there are a number of solutions based on bounding a design measure from above (e.g. a well-known exchange algorithm). In the first section of Chapter 5 an alternative method, that is based upon an extension of the concept of an information matrix, is presented and compared with standard techniques.

As for the second problem it is common that spatial phenomena exhibit strong local correlations, and thus classical methods of optimum design are applicable only under quite restrictive conditions. Thus, a separate design technique for correlated processes needs to be investigated. There have been several suggestions for the construction of optimum designs, which are briefly reviewed in the second section of Chapter 5 (for a detailed presentation see Näther (1985a)). The main part of Chapter 5 is concerned with adapting concepts from classical *optimum design theory*, such as design measures, to the correlated case. Here again the solution is based upon an extension or rather an approximation of the usual information matrix, that coincides with the classical one when no correlations arise, and that allows us to introduce design measures under correlation. This leads to the formulation of simple iterative algorithms for finding optimum or nearly optimum designs for estimating first-order characteristics of the spatial process. A method that can cope with measurements that are replicated in time is also presented. This is of importance, since spatial data (such as in environmental monitoring) often exhibit a spatio-temporal structure.

Multipurpose Designs Including Designs for Variogram Fitting

Chapter 6 discusses the applications of *optimum design theory* for estimation of second-order characteristics of spatial processes such as the variogram. Here similar problems as in Chapter 5 arise due to correlated observations. In addition one has to face the difficulty that the optimization is in the space of intersite distances. Thus another adaptation of design methods is needed to resolve the issues resulting from this property. Several design strategies are suggested and compared with respect to their performance. An extension of a well-known design algorithm for the correlated case from one to multiple iterative corrections is given.

It is also necessary to combine designs for estimation of first and second order characteristics efficiently. A method for combining several different purpose designs for good overall performance is proposed.

Example

In all chapters the formal descriptions are accompanied by an illustrating example taken from air-pollution monitoring, namely the reconstruction of the Upper-Austrian sulfuric dioxide (SO_2) monitoring network. This is a typical problem in spatial design and there exists a great number of similar applications in this field. A decisive aspect in this context is the scaling of the problem. Local phenomena (e.g. around industrial sites) strongly depend upon the physical (deterministic) description of the dispersion of emissions. On a larger scale (such as in the example given) we have transport effects that average out local influences, and a stochastic analysis is possible. For a description of practical problems related to the construction of monitoring networks, see Munn (1981).

The Upper-Austrian network consists of 17 observation sites (Lenzing, Linz—Hauserhof, Linz—Urfahr, Traun, Asten, Wels, Vöcklabruck, Perg, Steyr, Braunau, Linz—Kleinmünchen, Linz—Ursulinenhof, Linz-ORF-Zentrum, Linz—24er-Turm, Linz—Berufsschulzentrum, Steyregg-Weih, Schöneben), which are mainly concentrated at the capital and main industrial site Linz. A graphical representation of the observation sites on a 45 × 45 point grid is given in Figure 1.1. For computational reasons the map was rescaled to a unit square; one unit corresponds to approximately 83 kms.

Exercises

Every chapter (in its last section) contains a list of questions that relates to a second real example. The Südliche Tullnerfeld is a part of the Danube river basin in central Lower Austria and due to its homogeneous aquifer well suited for a model-oriented geostatistical analysis. It contains 36 official water quality measurement stations, which are irregularly spread over the region. A graphical representation of the sites (again after rescaling to a unit of approximately 31 kms) on a 95 × 95 grid is given in Figure 1.2. Suggested solutions to the exercises will appear, as they become available, on:

http://statistik.wu-wien.ac.at/stat4/mueller/csd/csd.htm

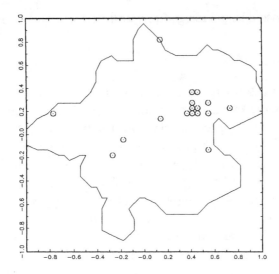

FIGURE 1.1. The Upper-Austrian SO$_2$ monitoring network; circles represent sites.

FIGURE 1.2. The water quality monitoring network in the Südliche Tullnerfeld; solid circles represent sites, grid points represent region.

Appendix

The example data set corresponding to the Upper-Austrian monitoring network contains 288 daily averages of half-hourly measured sulfuric dioxide (SO$_2$) concentrations in mg/m^3 taken over the period January 1994 until December 1995. Only days for which at least 40 half-hourly measurements were taken at each site are included.

The exercise data set contains measurements of chlorid (Cl) in mg/l over the period 1992–1997 on all time points, for which at least one measurement was taken (missing values are indicated by blanks).

An outline of the structure of both data sets is given in Appendix A.1. The full data sets can be downloaded from:

http://statistik.wu-wien.ac.at/stat4/mueller/csd/data.zip

To enhance readability of the text all original proofs of nontrivial mathematical facts from Chapters 2–6 are presented in Appendices A.2–A.6, respectively.

Most of the calculations in this monograph were carried out within the statistical programming environment GAUSS-386 V.3.2 (for syntax information see Aptech, Inc. (1993)). A so-called application module D2PT (two-dimensional design optimization) to facilitate (spatial) design calculation was developed. Its description can be found in Appendix A.7. A recent version of it can be downloaded from:

http://statistik.wu-wien.ac.at/stat4/mueller/csd/d2pt.zip

References

Aptech, Inc. (1993). *GAUSS-386 Command Reference*. Washington.

Arbia, G. (1989). *Spatial Data Configuration in Statistical Analysis of Regional Economic and Related Problems*. Kluwer Academic Publishers, Dordrecht.

Atkinson, A.C. and Fedorov, V.V. (1989). Optimum design of experiments. In S. Kotz, C.B. Read, N.L. Johnson, editors, *Encyclopedia of Statistical Sciences*, supplement volume. Wiley, New York, 107–114.

Bellhouse, D.R. (1988). Spatial sampling. In S. Kotz, C.B. Read, N.L. Johnson, editors, *Encyclopedia of Statistical Sciences*, volume 8. Wiley, New York, 581–584.

Christakos, G. (1992). *Random Field Models in Earth Sciences*. Academic Press.

Cressie, N. (1993). *Statistics for Spatial Data*, revised edition. John Wiley & Sons, New York.

Cox, D.D., Cox, L.H. and Ensore, K.B. (1997). Spatial sampling and the environment: some issues and directions. *Environmental and Ecological Statistics*, 4:219–233.

Fienberg, S.E. and Tanur, J.M. (1987). Experimental and sampling structures: Parallels diverging and meeting. *International Statistical Review*, 55:75–96.

Haining, R. (1990). *Spatial Data Analysis in the Social and Environmental Sciences*. Cambridge University Press, Cambridge.

Isaaks, E.H. and Srivastava, R.M. (1989). *An Introduction to Applied Geostatistics*. Oxford University Press, New York.

Matheron, G. (1963). Principles of geostatistics. *Economic Geology*, 58:1246–1266.

May, K. (1893). *Winnetou I*. Karl-May-Verlag, Bamberg.

Munn, R.E. (1981). *The Design of Air Quality Monitoring Networks*. MacMillan, London.

Näther, W. (1985). *Effective Observation of Random Fields*. Teubner Texte zur Mathematik — Band 72. Teubner Verlag, Leipzig.

2
Fundamentals of Spatial Statistics

"Space, the final frontier ..."
Roddenberry (1964)

In the past 20 years, there have appeared numerous monographs on spatial data analysis regarded from various aspects (e.g. Ripley (1981), Dutter (1985), Isaaks and Srivastava (1989), Wackernagel (1995), or Chiles, J. and Delfiner, P. (1999), amongst many others). Outstanding is the encyclopedic treatment by Cressie (1993). Due to the wide applications in geology a complete branch of statistics has been termed geostatistics and a very specific terminology has developed there. This chapter gives an overview over those geostatistical concepts that are relevant for the design issue and relates them to a classical statistics point of view.

Principally, there are two perspectives on the generation of (spatial) data: it is considered to be either completely deterministic or it contains a stochastic component. Let us for the moment accept the latter point of view and regard the data y observed at coordinates $x \in \mathcal{X} \subset \mathbb{R}^2$ as being generated by a parameterized process (random field)

$$y(x) = \eta(x, \beta) + \varepsilon(x), \qquad (2.1)$$

where η denotes the deterministic and ε the stochastic part. The compact set $\mathcal{X}$ is called the design space and β is assumed to stem from a finite dimensional parameter space $\Omega \subset \mathbb{R}^m$. Additionally to approximating the field by a parametric model $\eta(\cdot, \cdot)$, ε is usually assumed to have zero mean $E[\varepsilon(x)] = 0$, finite variances $E[\varepsilon(x)^2] = \sigma^2(x) < \infty$ and a (parameterized) covariance function

$$E[\varepsilon(x)\varepsilon(x')] = c(x, x'; \theta). \tag{2.2}$$

For simplicity we will only consider isotropic random fields, i.e. where the spatial correlation between two observations depends solely upon the distance between their respective sites, such that

$$c(x, x'; \theta) = c(h, \theta),$$

where $h = \|x - x'\|$, in a suitable norm. Generalizations of the concepts and methods discussed in this monograph to the anisotropic case are usually straightforward. Moreover, Sampson and Guttorp (1992) argue that in most cases a simple linear transformation of the coordinate system will restore isotropy.

A common interpretation is that the long range variability (mean, trend) in the field is generated by η, whereas the short range fluctuations are due to ε. We will thus denote them respectively as first and second order characteristics of the spatial process.

It is evident (also from the analogy to the similar problem in time-series analysis, see Brockwell and Davis (1991)) that there is no clear-cut distinction between first and second order characteristics. There exists an inherent impossibility to distinguish unequivocally the influences of local trends and spatial correlation. The more detailed the model for the trend is, the less of the systematic fluctuations of the field has to be ascribed to the spatial covariance model. The distinction is therefore a purely practical one and will be largely based upon interpretability and parsimony of the posited models.

In the stochastic framework (2.1) and (2.2) the main purpose of statistical analysis is then the estimation of β and θ and prediction of y based upon these parametric models. When $y(x)$ is considered to be deterministic, the aim is either interpolation to restore unobserved data or the calculation of overall field characteristics such as the mean. Since the former, the model-oriented, point of view is better embedded into traditional (mathematical) statistics, the contents of this monograph will be almost exclusively oriented towards it. Note, that the conclusions that can be reached under the two points of view can sometimes be confusing and contradicting. In restoring $y(\cdot)$ from

(2.1) it may be advantageous to have correlated errors and we therefore need less observations to reach a given precision than in the uncorrelated case (for examples see Section 5 in Haining (1990)). On the other hand correlated observations carry redundant information about η and we thus need more of them to estimate the model as precisely as from uncorrelated observations. For a comparison of the two standpoints and the resulting consequences for statistical analysis of spatial data, see Section 5.9 in Cressie (1993).

When it comes to the question of where to position observation sites most efficiently, the natural consequence of the model-oriented point of view is the application of results from *optimum design theory*, whereas adherents of the data-oriented deterministic approach are better served by *sampling theory* (e.g. Cassel *et al.* (1977)). Therefore, in Chapter 3 the main concepts from *optimum design theory* are reviewed for a better understanding of the subsequent chapters, where a special emphasis is put on the peculiarities of the spatial setting.

2.1 Estimation of Spatial Trend

Let us for a moment assume that c (including θ) from (2.2) is known. If we do not want to make further distributional assumptions it is natural that for estimation of the spatial trend $\eta(x,\beta)$ we are looking for the best linear unbiased estimator. Specifically we want to determine the values of β that minimize

$$\mathcal{S}[\beta, C] = [y - \eta(\beta)]^T C^{-1} [y - \eta(\beta)],$$

where $y^T = [y(x_1), \ldots, y(x_N)]$, $\eta^T(\beta) = [\eta(x_1, \beta), \ldots, \eta(x_N, \beta)]$, and $C_{ii'} = c(x_i, x_{i'}; \theta)$, $i, i' = 1, \ldots, N$, i.e. generalized least squares estimation.

If we apply a Taylor expansion based linearization of η in a Gauss-Newton algorithm we arrive at

$$\hat{\beta} = [\dot{\eta}^T(\hat{\beta}) C^{-1} \dot{\eta}(\hat{\beta})]^{-1} \dot{\eta}^T(\hat{\beta}) C^{-1} y, \qquad (2.3)$$

where $\dot{\eta}(x, \hat{\beta}) = \frac{\partial \eta(x, \beta)}{\partial \beta}|_{\beta=\hat{\beta}}$, and $\dot{\eta}^T(\hat{\beta}) = [\dot{\eta}(x_1, \hat{\beta}), \ldots, \dot{\eta}(x_N, \hat{\beta})]$, under the usual regularity conditions from nonlinear least squares theory. Note that this estimator, although unbiased in the linearized setting is biased in the original setting, see e.g. Pázman (1993).

In geostatistical practice of course the assumption of known C is unrealistic and usually we assume knowledge of it only up to unknown parameters. It is then convenient to adopt iteratively reweighted least squares estimation:

$$\hat{\beta}^{(s+1)} = [\dot{\eta}(\hat{\beta}^{(s)})C^{-1}(\hat{\theta}^{(s)})\dot{\eta}^T(\hat{\beta}^{(s)})]^{-1}\dot{\eta}(\hat{\beta}^{(s)})C^{-1}(\hat{\theta}^{(s)})y,$$

$$\tilde{\beta} = \lim_{s\to\infty}\hat{\beta}^{(s)}, \quad \tilde{\theta} = \lim_{s\to\infty}\hat{\theta}^{(s)} \qquad (2.4)$$

where $\theta^{(s)}$ stems from a fit of $C(\hat{\theta}^{(s)})$ to $\frac{1}{N-m}[y-\eta(\hat{\beta}^{(s)})][y-\eta(\hat{\beta}^{(s)})]^T$ at every step s by a suitable technique (see section 2.5). In practice the iterations have to be started at an initial guess (say $\hat{\beta}_0 = \hat{\beta}_{OLS}$). The procedure then generally yields asymptotically efficient and consistent estimates. For an overview of such properties and further details see del Pino (1989). The resulting estimate of the asymptotic dispersion matrix of $\tilde{\beta}$ is then consequently

$$\widetilde{\text{Cov}}[\tilde{\beta}] = [\dot{\eta}(\tilde{\beta})C^{-1}(\tilde{\theta})\dot{\eta}^T(\tilde{\beta})]^{-1}.$$

Note that $\{\tilde{\beta},\tilde{\theta}\}$ is a consistent estimator. However, it does not generally minimize $S[\beta, C(\theta)]$ (cf. Fedorov (1972), page 45); this is a surprising fact that proves relevant for variogram fitting (see Section 2.5).

2.2 Universal Kriging

Having estimated the spatial trend by $\eta(x,\tilde{\beta})$ we can now turn our attention to the problem of prediction of unobserved data $y(x_0)$ at a prespecified site x_0. A natural predictor is

$$\hat{y}(x_0) = \eta(x_0,\tilde{\beta}) + l_0^T\hat{\varepsilon}, \qquad (2.5)$$

with

$$l_0 = \arg\min_l E[\tilde{\varepsilon}(x_0) - l^T\hat{\varepsilon}]^2, \qquad (2.6)$$

where $\hat{\varepsilon}^T = [\hat{\varepsilon}(x_1),\ldots,\hat{\varepsilon}(x_N)]$ is the residual vector and $\tilde{\varepsilon}(x_0) = y(x_0) - \eta(x_0,\tilde{\beta})$. The rule given by (2.6) means minimizing the square risk of the predictor $\hat{y}(x_0)$.

Unfortunately, (2.6) has no unique solution, since $E[\hat{\varepsilon}\hat{\varepsilon}^T]$ is singular, however one admissible solution (see Fedorov (1989)) is

$$l_0 = C^{-1}c(x_0),$$

with $c^T(x_0) = [c(x_0, x_1; \theta), \ldots, c(x_0, x_N; \theta)]$. This, naturally combined with (2.5) yields the predictor

$$\hat{y}(x_0) = \eta(x_0, \tilde{\beta}) + c^T(x_0) C^{-1} [y - \eta(\tilde{\beta})]. \qquad (2.7)$$

For the case of a linear response, i.e $\eta(\beta) = [x_1, \ldots, x_N]^T \beta = X\beta$, and by substituting the standard linear generalized least squares estimator, we can rewrite (2.7) as

$$\hat{y}(x_0) = [c(x_0) + X(X^T C^{-1} X)^{-1}(x_0 - X^T C^{-1} c(x_0))]^T C^{-1} y, \qquad (2.8)$$

which is the well-known universal kriging estimator (Goldberger (1962), see Section 3.4.5 in Cressie (1993)). The corresponding variance of the predictor (the kriging variance) is given by

$$\begin{aligned}\text{Var}[\hat{y}(x_0)] &= \sigma^2(x_0) - c^T(x_0) C^{-1} c(x_0) + \\ &\quad [x_0 - X^T C^{-1} c(x_0)]^T (X^T C^{-1} X)^{-1} [x_0 - X^T C^{-1} c(x_0)].\end{aligned} \qquad (2.9)$$

Note that X may not only contain the regressor variables but by a corresponding modification of (2.8) and (2.9) also functions thereof, e.g. squared or interaction terms in order to model second degree polynomial surfaces.

In the light of the discussion so far universal kriging can be viewed as a two stage procedure involving GLS estimation of the trend surface and best linear prediction, as pointed out by Fedorov (1989). For a reverse two-stage interpretation see again Cressie (1993). This view allows us to recognize the fundamental problem connected with this prediction technique: it strictly retains its optimality property (2.6) only if (either $\eta(\beta)$ or) as above C is known, which is hardly the case in the real world. Usually C is replaced by an empirical estimate and the effect of this practice on the quality of the kriging estimator is not very well understood (especially in small samples). Asymptotically, however, the effects of misspecifying the covariance function seem negligible as long as its functional form is compatible (see Stein (1988)).

Application of Universal Kriging

Let us, in spite of the disadvantages reported in the first part of this section, take a look at the current geostatistical practice: In a first stage the spatial trend is eliminated from the data. It is either assumed to be known from external sources or estimated by means of methods from Section 2.2 or similar.

In geostatistical data analysis second order spatial dependence is usually not characterized by the covariance function c but by the so-called variogram

$$2\gamma(h) = E\{[y(x') - y(x)]^2\} \qquad \forall x, x' \in \mathcal{X} \subset \mathbb{R}^2, h = \|x - x'\|,$$

which is assumed to have a known parametric form, so that we have (after detrending) the properties

$$E\{[y(x_i) - y(x_{i'})]\} = 0,$$

and

$$E\{[y(x_i) - y(x_{i'})]^2\} = 2\gamma(h_{ii'}, \theta),$$

where $h_{ii'} = \|x_i - x_{i'}\|$, i.e. observations from the so-called intrinsically stationary random field. This parametric form is fitted to the residuals (or more preferably a learning sample) to yield $\hat{\theta}$, e.g. by means described in Section 2.5.

It has to be noted that if the covariance function exists there is a strict relationship between it and the variogram, which here simply is

$$2\gamma(h, \theta) = \sigma^2(x) + \sigma^2(x') - 2c(h, \theta'),$$

where θ' represents a specific parameterization of the covariance function, which is in one-to-one correspondence with the variogram's parameterization by θ. Utilizing this property it is possible to derive from (2.8) a dual formulation of the (empirical) universal kriging estimator

$$\widehat{y}(x_0) = [\hat{\gamma}(x_0) + X(X^T\hat{\Gamma}^{-1}X)^{-1}(x_0 - X^T\hat{\Gamma}^{-1}\hat{\gamma}(x_0))]^T\hat{\Gamma}^{-1}y, \quad (2.10)$$

where $\hat{\gamma}^T(x_0) = [\gamma(h_{01}, \hat{\theta}), \ldots, \gamma(h_{0N}, \hat{\theta})]$ and $\hat{\Gamma}_{ii'} = \gamma(h_{ii'}, \hat{\theta}); i, i' = 1, \ldots, N$. Note, that the inverse of $\hat{\Gamma}$ in (2.10) exists, when some of the easily fulfilled properties in Section 2.4 hold.

It must once more be emphasized that not only is $\widehat{y}(x_0)$ biased but also its variance may be considerably larger than (2.9) due to the replacement of γ by $\hat{\gamma}$, particularly if the same set of data is used for both estimation stages. For taking this into account Zimmerman and Cressie (1992) derive approximations to the 'true' mean squared error under rather restrictive assumptions. Pilz et al. (1997) try to overcome the problem by finding the predictor with the minimum possible kriging variance with respect to a whole class of plausible variogram models, a method they term *minimax kriging*. A general discussion of the adverse effects of variogram fitting on the quality of the kriging predictor can be found in Cressie (1993), Section 5.3.

2.3 Local Regression

The described drawbacks of universal kriging lead to the consideration of alternative prediction techniques. One possibility is to estimate the trend surface by nonparametric regression, which gives us enough flexibility to absorb the signal almost completely into the trend. We then arrive at (approximately) independent residuals, which radically facilitates statistical analysis and design considerations.

Let us therefore in the spirit of Hastie and Tibshirani (1993) consider relationships of the form

$$y(x) = \eta(x, \beta(x)) + \varepsilon(x), \qquad (2.11)$$

where η is assumed to be a smooth function of x and now $E[\varepsilon(x)\varepsilon(x')] = \sigma^2(x)\delta_{x,x'}$. By allowing β to change smoothly over the location x we may yield consistent estimates $\hat{y}$ from the data (X, y).

A natural choice for an estimation procedure for $\beta(\cdot)$ in the setting (2.11) is weighted least squares, where the weights w are made dependent upon the distances between sites x_i and x_0:

$$\hat{\beta}(x_0) = \arg\min_{\beta(x_0)} \sum_{i=1}^{N} w(x_i, x_0) \left[y_i - \eta\left(x_i, \beta(x_0)\right)\right]^2, \qquad (2.12)$$

and consequently

$$\hat{y}(x_0) = \eta(x_0, \hat{\beta}(x_0)).$$

This procedure is called local regression and was introduced to the statistical literature by Cleveland (1979), whose primary intention was to find a tool that enhances the visual information to be extracted from a scatterplot. However, the method was used before in the geosciences and it was probably Pelto *et al.* (1968), who firstly applied it for producing contour maps from irregularly spaced data. Its wider use in spatial statistics has been advocated by Ripley (1981) and Fedorov (1989). A recent, more complete survey of the method and its applications can be found in Cleveland and Loader (1996).

Local Response

The local regression method relies heavily on the idea that a sufficiently smooth function can always be approximated by a 'simpler' function over a small region of the regressor space. Thus each local response can also be interpreted as a (two-dimensional) Taylor expansion of the true response up to an order p at the point of interest, i.e.

$$\eta\left(x_i, \beta(x_0)\right) = \sum_{k+j \leq p} \beta_{jk}(x_0)(x_{i1}-x_{01})^j(x_{i2}-x_{02})^k + r_i, \quad i=1,\ldots,N,$$

where r_i is the remainder term of the approximation, vanishing as $o(h_i^p)$ when $x_i \to x_0$ and thus

$$\hat{y}(x_0) = \hat{\beta}_{00}(x_0). \tag{2.13}$$

Note that for isotropic fields an alternative and simpler local response may be used, namely

$$\eta\left(x_i, \beta(x_0)\right) = \sum_{j=0}^{p} \beta_j(x_0) h_i^j + r_i, \quad i=1,\ldots,N, \tag{2.14}$$

where $h_i = \|x_0 - x_i\|$.

Obviously the choice of the order p of the local polynomial is somewhat arbitrary and there exists some freedom in choosing the degree of the local approximation. The point is that insufficient local fitting may introduce a bias to the estimator, especially if the curvature of the original surface is high. It is clear, that through introducing more sophisticated local fitting schemes it will be possible to absorb the bias effect, though of course the computational burden may increase tremendously. Fedorov et al. (1993) have investigated the behavior of the estimator when the local response is misspecified, a corresponding simulation experiment can be found in Maderbacher and Müller (1995). Those results indicate that a good compromise between computational ease and flexibility is the choice $p=1$, i.e. a local linear fit.

Weights

The weight function has to be constructed in a way that the influence of an observation diminishes with increasing distance of its respective collection site x_i to a point of interest x_0. It is therefore practical to make the weight functions involved in (2.12) dependent upon the distance $h_i = \|x_i - x_0\|$, following the properties below:

2.3 Local Regression

- nonnegative: $w(h_i) \geq 0$,

- nonincreasing: $w(h_i + \epsilon) \leq w(h_i)$, $\forall \epsilon > 0$,

- possibly vanishing: $w(h_i) = 0$, $\forall h_i > \kappa(x_0) < \infty$,

or at least local (see Appendix A.2.2): $\lim_{h_i \to \infty} h_i^2 w(h_i) = 0$.

For the vanishing property there are two distinct choices. We can either hold $\kappa(x_0) = \kappa < \infty$ fixed or we choose it such that a fixed size neighborhood of x_0 is included, i.e. $\kappa(x_0)$ is equal to the distance of the αN-th nearest neighbor of x_0 amongst the x_i's, where $0 < \alpha \leq 1$. We will refer in what follows to the fixed or αN-th nearest neighborhood method, respectively.

It is also advisable to use a weight function that decreases smoothly to 0 as h goes to κ. It will help in producing $\hat{y}_i$'s, that have a reasonably smooth appearance. For this reason uniform weights are generally not a good choice. The value of $\kappa(\cdot)$ largely determines the amount of smoothness of the resulting estimator and is thus usually referred to as the smoothing parameter.

One particular function that has the above desired properties and is therefore often used by Cleveland (1979) and his followers is the so-called tricube:

$$w(h) = \begin{cases} \left(1 - \left(\frac{h}{\kappa(x_0)}\right)^3\right)^3, & 0 \leq h \leq \kappa(x_0) \\ 0 & \text{else} \end{cases} \quad (2.15)$$

An example of a weight function for a fixed neighborhood is

$$w(h) = \frac{e^{\frac{-h^2}{\kappa^2}}}{h^2 + \epsilon}, \quad (2.16)$$

which was suggested out of practical considerations by McLain (1971). Here κ is chosen as the average distance of the data points from the points of interest and ϵ is a small number introduced for numerical stability. Figure 2.1 gives a graphical representation for the weight functions (2.15) and (2.16) for $\kappa(\cdot) = \kappa$.

Note that for all weight functions κ has to be selected in a way that guarantees that at least $p + 1$ (for a local linear fit, or otherwise respectively more) points are included in each neighborhood.

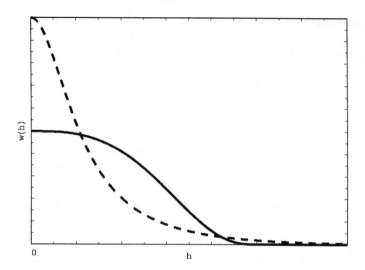

FIGURE 2.1. A generic tricube (solid) and McLain's (dashed) weight function.

Amount of Smoothness

It is evident that the size of the neighborhood of a point x_0 determines the amount of how much the original series is smoothed. The required smoothing is strongly dependent upon the characteristics of the random field. However, there are several recommendations as for how to select the smoothing parameter κ (or α). One of the most natural approaches is cross validation (cf. Stone (1974)), i.e. to choose

$$\kappa^* = \arg\min_{\kappa} \sum_{i=1}^{N} \left(y_i - \hat{y}_{(i)}(\kappa)\right)^2, \qquad (2.17)$$

where $\hat{y}_{(i)}$ denotes the local fit at x_i, when y_i is not used in the computation. A comparison of the properties of various techniques including (2.17) can be found in Härdle et al. (1988). Droge (1996) presents an alternative global procedure, whereas Rajagopalan and Lall (1998), in a spatial context, employ a localized cross validation technique.

Statistical Properties

The estimator (2.13) applied to all points in the sample can be rewritten as a linear estimator

$$\hat{y} = Ly.$$

Note that L is neither symmetric nor idempotent as in the linear, parametric case. If the smoothing parameter and the local response are chosen properly we can be confident of unbiasedness of the estimator with resulting variance-covariance matrices

$$\mathrm{Cov}(\hat{y}) = \sigma^2 L L^T \qquad (2.18)$$

and

$$\mathrm{Cov}(\hat{\varepsilon}) = \sigma^2 (I - L)(I - L)^T = \sigma^2 F.$$

It is a remarkable feature that (2.18) is usually a nonincreasing function of smoothing parameters. That means that making the estimates smoother tends to increase their bias and reduce their variances, whilst making them more unsmooth has the opposite effect.

If we now suppose, that the smoothness is such that the bias in the fitted values is negligible, the error variance can be estimated by

$$\hat{\sigma}^2 = \frac{\sum_{t=1}^n \hat{\varepsilon}_t^2}{\mathrm{tr}(I - L)(I - L)^T} = \frac{\hat{\varepsilon}^T \hat{\varepsilon}}{\mathrm{tr} F}$$

and under the assumption of identically distributed, independent Gaussian noise $\frac{\hat{\sigma}^2 (\mathrm{tr} F)^2}{\sigma^2 \mathrm{tr} F^2}$ is approximately distributed as a χ^2 with $l.d.f.$ (lookup) degrees of freedom, where $l.d.f.$ is $\frac{(\mathrm{tr} F)^2}{\mathrm{tr} F^2}$ rounded to the nearest integer. Note, that this approximation gets better (up to a third moment match) if the weight function (2.15) is chosen, as is argued by Cleveland (1979).

Prediction

Since under the assumptions of the previous section the linearity of the estimator holds at every point x_0, we have

$$\hat{y}(x_0) = \eta\left(x_0, \hat{\beta}(x_0)\right) = L(x_0) y$$

and its variance is therefore

$$\mathrm{Var}(\hat{y}(x_0)) = \sigma^2 L(x_0) L^T(x_0), \qquad (2.19)$$

which can be estimated by $\hat{\sigma}^2(x_0) = \hat{\sigma}^2 L(x_0) L^T(x_0)$ and thus the random variable $\frac{\hat{y}(x_0) - \eta(x_0, \beta(x_0))}{\hat{\sigma}(x_0)}$ is approximately t-distributed with $l.d.f.$ degrees of freedom for normal errors.

Equivalent Number of Parameters

The quantity
$$e.n.p. = N - \mathrm{tr} F \qquad (2.20)$$
can be interpreted as an equivalent number of parameters, since for an unweighted least squares fit of a linear model it gives the number of parameters $p+1$. It thus can be used to assess the amount of smoothness under different conditions and regimes. For local polynomial smoothing (2.14) and the tricube weight function (2.15) Cleveland et al. (1992) give an approximate relation between $e.n.p.$ and the smoothing parameter α as $\alpha \simeq 1.2 \frac{p}{e.n.p}$. Analogously, $N - \mathrm{tr} LL^T$ is termed equivalent degrees of freedom ($e.d.f.$). However, it is not generally the same as $l.d.f.$, the lookup degrees of freedom, as in parametric regression. For a broad discussion of these and other concepts of smoothness evaluation see Buja et al. (1989).

Other Smoothers and Nonparametric Kriging

There exists a vast number of smoothing methods that could alternatively be used in (2.11), such as kernel smoothing, k-nearest neighbor estimates, spline smoothing, median polishing, to mention just a few. An excellent survey with emphasis on kernel smoothing gives Härdle (1990). Other comprehensive expositions are the monographs by Eubank (1988) and Hastie and Tibshirani (1990).

However, in Fan (1993) we can find a strong argument in favor of local linear smoothers as opposed to those alternatives. If one considers only regression functions with finite second derivatives $\{\eta(\cdot, \beta) : |\frac{d^2\eta(x,\beta)}{dx^2}| \leq \kappa < \infty\}$, then the best (according to its minimax risk) among all linear smoothers is the local linear smoother. Its major disadvantage, the loss of statistical power when the data gets sparse as the dimensionality of $\mathcal{X}$ grows (the so-called curse of dimensionality) is of less importance in the spatial context where the dimension is usually not larger than three.

Principally, there is no reason not to exploit potential information remaining in the residuals after a smoothing step. The procedure is similar to the second step in universal kriging. For local regression this was applied in a series of papers started by Haas (1990) with an extension to spatio-temporal structures by Haas (1995). For a detailed discussion of the techniques with median polishing as the smoother see Section 3.5 in Cressie (1993). There have been suggestions on explicitly correcting the smoother for correlated errors, see e.g. Altman (1990) or Opsomer (1996). A unified approach that encompasses both kriging and local regression as special cases is provided by Host (1999).

2.4 Variogram Fitting

For effective prediction in correlated random fields it is necessary to obtain good estimates of the second order properties, see Guttorp and Sampson (1994). Specifically, for the universal kriging predictor (2.10) we require the (parametric) estimation of the variogram. It is natural to initially estimate it nonparameterically by

$$2\hat{\gamma}_k = \frac{1}{N_{H_k}} \sum_{x_i, x_{i'} \in H_k} [y(x_i) - y(x_{i'})]^2. \qquad (2.21)$$

For the classical variogram estimator by Matheron (1963) the set (in this context sometimes called bin) H_k contains all point pairs $x_i, x_{i'}$ with distances such that $h_{ii'} = h_k$ (N_{H_k} denotes their respective number). The sequence h_k, $k = 1, \ldots, K$ consists of all ordered distinct distances. If the data stem from an irregular grid this set is usually enlarged to contain a certain neighborhood of k, but always such that $\sum_{k=1}^{K} N_{H_k} = \binom{N}{2}$. A parametric fit $\gamma(h, \hat{\theta})$ of the (semi-)variogram is usually found by least squares regression through the points $(h_k, \hat{\gamma}_k)$.

For the proper use of a parameterized variogram $\gamma(h, \theta)$ in universal kriging (e.g. to guarantee positive definiteness of $\hat{\Gamma}$ or $\hat{C}$) it is required to have a number of properties. Such a model is termed a valid variogram model and it possesses certain attributes that are related to the spatial dependence structure of the process and that can therefore have a bearing on design. In our simplified context the main features (for any $\theta \in \Theta$) are

- $\gamma(\cdot)$ is negative semidefinite, i.e., $\sum_i \sum_j \lambda_i \lambda_j \gamma(h_{ij}, \theta) \leq 0$ for all $\lambda_1, \lambda_2, \ldots$ such that $\sum_i \lambda_i = 0$,

- $\gamma(0, \theta) = 0$, and $\lim_{h \to 0} \gamma(h, \theta) = \epsilon \geq 0$, i.e. a possible discontinuity at the origin, the so-called 'nugget effect', which allows for replicated measurements at a specific point,

- $\gamma(h, \theta)$ is smoothly increasing as h increases, up to

- a so-called 'range' a, such that $\gamma(h,\theta) = \max_{h'} \gamma(h',\theta)$ for all $h \geq a$, the so-called 'sill'.

A complete description of these conditions (and less restrictive variants) can be found in Journel and Huijbregts (1978), pp. 36-40, 163-183.

Perhaps the most widely used parametric variogram model that satisfies the above conditions is the so-called spherical variogram

$$\gamma_S(h,\theta) = \begin{cases} 0, & h = 0 \\ \theta_1 + \theta_2 \cdot \left\{ \frac{3}{2}\left(\frac{h}{\theta_3}\right) - \frac{1}{2}\left(\frac{h}{\theta_3}\right)^3 \right\}, & 0 < h \leq \theta_3 \\ \theta_1 + \theta_2, & h > \theta_3 \end{cases}.$$

This variogram increases monotonically from a 'nugget effect' of θ_1 near the origin to a 'sill value' of $\theta_1 + \theta_2$, which is attained at the 'range' θ_3, see Fig 2.2. The respective parameter space is $\Theta = \{\theta : \theta_1 \geq 0, \theta_2 > 0, \theta_3 > 0\}$.

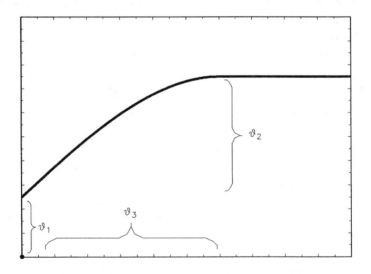

FIGURE 2.2. A generic spherical variogram (h horizontal, γ vertical).

Estimation from the Variogram Cloud

One particular case of (2.21) is to choose each H_k to contain only one element (pair) with $h_{ii'}$, i.e. $N_{H_k} \equiv 1$, $k = 1, \ldots, \binom{N}{2} \equiv K$. The collection of pairs $(h_{ij}, \hat{\gamma}_k)$ is then called the (semi-) variogram cloud and we attempt to estimate $\gamma(h, \hat{\theta})$ directly from the (semi-)variogram cloud without prior binning and averaging. Note that although this is a deviation from standard geostatistical practice, it clearly utilizes the maximum available information and is thus, if feasible, preferable. This is confirmed by simulation results in Müller (1999).

If we now assume that $y(\cdot)$ is a Gaussian random field, Cressie (1985) showed that $E(\hat{\gamma}_k) = \gamma(h_k)$, and the corresponding covariance matrix has entries

$$\Sigma_{kl} = \text{Cov}(\hat{\gamma}_k, \hat{\gamma}_l) \tag{2.22}$$
$$= \frac{1}{2} \left[\gamma(\|x_i - x_j\|) + \gamma(\|x_{i'} - x_{j'}\|) - \gamma(\|x_i - x_{j'}\|) - \gamma(\|x_{i'} - x_j\|) \right]^2,$$

where $\|x_j - x_{j'}\| = h_l$. Thus we have $\text{Var}(\hat{\gamma}_k) = 2\gamma^2(h_k)$.

Clearly from (2.22), the 'observations' $\hat{\gamma}_k$ are generally correlated and thus the ordinary least squares procedure yields inefficient estimates. The situation would require generalized least squares estimation, however, for computational simplicity Cressie (1985) recommended to use the following (weighted least squares) estimator based on the simplifying assumption that $\Sigma \simeq \text{diag}(2\gamma^2(\theta))$:

$$\tilde{\theta} = \arg\min_\theta \sum_{k=1}^{K} \gamma^{-2}(h_k, \theta) [\hat{\gamma}_k - \gamma(h_k, \theta)]^2. \tag{2.23}$$

It has to be noted that for variograms with finite range the assumption of approximately zero off-diagonal elements of Σ can be fulfilled by a proper choice of the sampling sites (for a respective algorithm see Section 6.3) and a corresponding deletion of large distances.

Estimator $\tilde{\theta}$ became one of the most widely used in variogram estimation (see e.g. Gotway (1990), Cressie (1993), Lamorey and Jacobson (1995)). However, several authors have noticed poor finite sample behavior (see e.g. McBratney and Webster (1986) and Zhang et al. (1995)). Simulation studies by Zhang et al. (1995) and Müller (1999) back up this claim.

It is natural to employ asymptotic theory to investigate possible causes of this deficiency. If $\tilde{\theta}$ is found by direct minimization of (2.23) from the (semi-)variogram cloud there is strong indication that it is a generally inconsistent estimator (see Appendix A.2.1. and similar types of estimators in Fedorov (1974)). This is not the case under the assumptions A1 (K fixed) and A2 ($N_{H_k} \to \infty$) of Cressie (1985), but here we have $N \to \infty$ and thus $K \to \infty$ and $N_{H_k} = 1 \quad \forall k$. Moreover, the small sample simulation results from Müller (1999) show that the studied asymptotics may have some explanatory value also for the binned estimator, especially when the number of bins is large.

A consistent estimator can also be found by utilizing iteratively reweighted least squares or the slightly simpler updating scheme from Seber and Wild (1989), Section 2.8.8,

$$\hat{\theta}^{(s+1)} = \hat{\theta}^{(s)}$$
$$+ \{\dot{\gamma}(\hat{\theta}^{(s)})^T \Sigma^{-1}[\gamma(\hat{\theta}^{(s)})]\dot{\gamma}(\hat{\theta}^{(s)})\}^{-1} \dot{\gamma}(\hat{\theta}^{(s)})^T \Sigma^{-1}[\gamma(\hat{\theta}^{(s)})][\hat{\gamma} - \gamma(\hat{\theta}^{(s)})],$$
$$\bar{\theta} = \lim_{s \to \infty} \hat{\theta}^{(s)} \tag{2.24}$$

where $\dot{\gamma}(\hat{\theta}) = \frac{\partial \gamma}{\partial \theta}|_{\theta=\hat{\theta}}$ and the entries of $\Sigma[\gamma(\hat{\theta}^{(s)})]$ can be calculated from a parametric version of (2.22). This algorithm is computationally much more intensive than (2.23) and requires the inversion of a $K \times K$ matrix at each step. This estimator has been suggested in this context already by Cressie (1985) and used by various authors without justifying the cause (e.g. McBratney and Webster (1986), Zimmerman and Zimmerman (1991), and Genton (1998)). It was again Fedorov (1989), who correctly argued for and described an iterative algorithm for the case $\Sigma = \text{diag}(2\gamma^2(\theta))$.

A framework, in which the second order characteristics of the response is estimated by local regression (similar to Section 2.3 for trend estimation) is provided by Ruppert et al. (1997).

2.5 Example

Preliminary Analysis — Detrending and Data Transformation

Like many other data sets in environmental statistics the SO_2 measurement data from the Upper-Austrian air-pollution monitoring network are not only attached to two spatial coordinates but also to a temporal index. This is usually not the case for 'classical' geostatistical data such as mining results from drilling holes. To cope with this particular feature it is necessary to extend the basic model (2.1), e.g. by

$$y_t(x) = \eta(x, t; \beta) + \varepsilon_t(x), \qquad t = 1, \ldots, T, \tag{2.25}$$

with a spatio-temporal covariance structure for ε_t.

Since the data set is used in an illustrative way and the main focus of the monograph is on the spatial component we will simply assume that observations from each day represent independent replicates of the spatial field, i.e.

$$\text{Cov}(y_t(x), y_{t'}(x')) = \delta_{t,t'} c(h, \theta). \qquad (2.26)$$

However, this assumption leads us to a simple estimator of the spatial trend — the temporal average. It allows us to detrend the data without making assumptions about the form of the surface and prevents us from the pitfalls introduced by the techniques in Section 2.2. After this detrending the explorative methodology by Box and Cox (1964) was used to identify the square-root transformation for removing the apparent skewness. The location coordinates, the two-year average $\bar{y}(x_i)$ and the detrended and transformed observations $\tilde{y}_t(x_i) = \pm\sqrt{y_t(x_i) - \bar{y}(x_i)}$ for a typical day ($t = 17$, 94-03-01) are given in Table 2.1.

Station name	$x_{i[1]}$	$x_{i[2]}$	$\bar{y}(x_i)$	$\tilde{y}_{17}(x_i)$
Linz—Hauserhof	0.4112	0.2348	8.68	1.56
Linz—Urfahr	0.4112	0.3524	5.23	1.93
Traun	0.3524	0.1760	5.72	0.85
Asten	0.5288	0.1760	4.75	0.55
Wels	0.1172	0.1172	6.43	0.91
Vöcklabruck	-0.1768	-0.0592	4.35	0.45
Perg	0.7052	0.2348	4.72	0.53
Steyr	0.5288	-0.1180	4.68	0.34
Braunau	-0.7648	0.1760	3.83	-0.76
Linz—Kleinmünchen	0.4112	0.1760	4.72	1.16
Linz—Ursulinenhof	0.4112	0.2936	8.36	2.11
Linz—ORF-Zentrum	0.4700	0.2348	8.24	1.52
Linz—24er-Turm	0.4700	0.3524	5.44	2.13
Linz—Berufsschulzentrum	0.4700	0.1760	7.56	1.06
Steyregg-Weih	0.5288	0.2936	6.92	0.47
Lenzing	-0.2944	-0.1768	8.39	-0.15
Schöneben	0.1172	0.8228	4.05	-0.06

TABLE 2.1. Monitoring station coordinates, two-year averages and transformed observations from a typical day.

It must be stressed that the above assumption of temporal independence is a very rough one, as can be easily seen by looking at a

26   2. Fundamentals of Spatial Statistics

temporal plot of the data for a typical site, e.g. Figure 2.3. It is apparent that there is a considerable increase in SO_2 during the winter months, however, its size is relatively small as compared to the amount of fluctuation in the data. Thus, the approximation above is to some extent justified and there will be only minimum exploitation of the temporal nature of the data in the rest of the monograph.

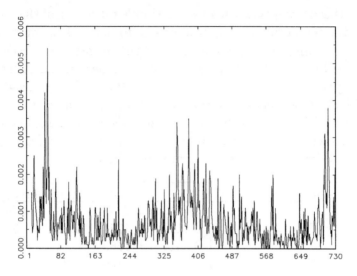

FIGURE 2.3. Daily averages of SO_2 concentrations in mg/m^3 in Steyregg-Weih (horizontal time, vertical SO_2).

There are other dimensions of the data set that are not considered fully in this monograph: differing atmospheric patterns may cause differing trend structures, predominant wind directions are likely to introduce anisotropy, the topography can affect the transportation of pollutants, etc. A discussion of these 'exogenous' influences and their statistical treatment can be found in Thiebaux and Pedder (1987).

Local Regression

A contour plot of the mean surface was estimated by local regression techniques, see Figure 2.4. Cross validation (2.17) over a fixed size neighborhood and using the tricube weight function (2.15) yielded $\kappa = 1.721$ with a respective $\hat{\sigma}_0 = 2.899$. This resulted in corresponding $e.n.p. \simeq 4$, i.e. slightly less smooth than a plane. The respective degrees of freedom were $e.d.f. = 13.8$ and $l.d.f. = 15$.

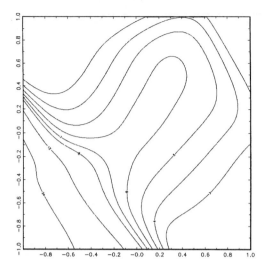

FIGURE 2.4. Contour plot of a local regression estimate of the trend surface.

Variogram Fitting

Figure 2.5 represents four typical cases of relative behavior of the variogram estimates. In all graphs the solid line gives the classical variogram estimate ((2.21) with $K = 7$), the dotted line is the OLS estimate (i.e. (2.23) without weighting), the dashed line stands for $\gamma(h_k, \bar{\theta})$ from (2.24) and the closely dashed for $\gamma(h_k, \tilde{\theta})$ from (2.23), the latter three estimated from the variogram cloud.

It is obvious from the graphs that $\gamma(h_k, \tilde{\theta})$ is considerably biased also in small samples, which can be ameliorated by applying the correction factor $\frac{1}{3}$ (see Appendix A.2.1). The most frequently observed case is as on day 94-03-01 (upper right), where the OLS and the iteratively reweighted estimate nearly coincide. On the other hand quite a few cases look like day 94-03-16 or 94-12-16 (lower left and lower right), where there is a considerable discrepancy between those two estimates, either in the sill or in the range. Rarely situations like 94-02-09 (upper left) occur, where $\gamma(h_k, \bar{\theta})$ lies even above $\gamma(h_k, \tilde{\theta})$.

Kriging

For detrended observations (2.10) reduces to

$$\hat{y}(x_0) = \hat{\gamma}^T \hat{\Gamma} y,$$

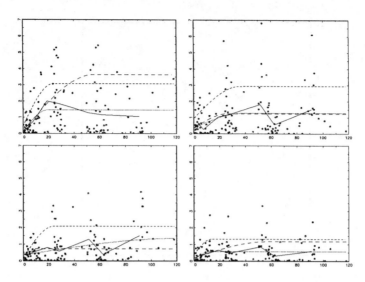

FIGURE 2.5. Variogram cloud (dots) and various variogram estimators for the days 94-02-09, 94-03-01, 94-03-16, 94-12-16 (from upper right to lower left, vertical γ, horizontal h).

which is known as simple kriging. We utilize the variogram parameters $\tilde{\theta}^T = (0.164, 0.323, 24.25)$ obtained in the previous section for 94–03–01. The contour plot for the square of the kriged surface for the detrended and transformed data is given in Figure 2.6. It can thus be added to the contour plot of Figure 2.4 to yield a nonparametric kriging estimator similar to the ones presented in Haas (1990).

2.6 Exercises

Table 2.2 gives the station coordinates and the temporal average of chlorid (Cl) concentrations for the respective station and the period 1992–1997 in the Südliche Tullnerfeld.

1. Select a station and check graphically whether the temporal mean can be considered a meaningful estimate of the spatial trend. If not try to divide up the periods to make it so.

2. Check whether there is skewness in the data and remove it by an appropriate transformation.

3. The data is suspect to anisotropy since there is a dominant flow direction (from west to east) in the aquifer. Guttorp and

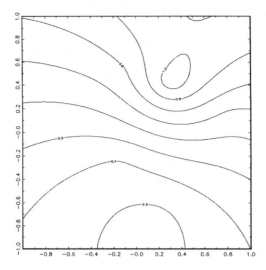

FIGURE 2.6. Squared simple kriging contour plot of the detrended data.

Sampson (1994) suggest to fit the variogram after a linear coordinate transformation $x \to Qx$. Select a particular day and estimate the appropriate 2×2 transformation matrix $\hat{Q}$ by introducing a separate step in (2.24). Compare results for various days for consistency.

4. In the transformed coordinate system perform a local linear regression fit for the mean surface by selecting the smoothing parameter by cross validation. Can you take account of the differing number of nonmissing observations N_T? If the result is not satisfactory try a different local model.

5. Construct a nonparametric kriging contour plot by adding simple kriging predictions to the smoothed surface. For comparison give a graphical representation of the transformed design region and site coordinates.

Connections to Other Chapters

In this chapter the basic geostatistical prediction tool universal kriging was presented in the framework of generalized least squares estimation. The interconnected problem of variogram estimation was emphasized. Additionally, local regression was shown to be an alternative or complementary method for spatial prediction. In Chapter 3 the fundamentals of optimum experimental design will be given, where the concentration is on regression functions with uncorrelated errors, the more general correlated case that corresponds to the basic random field model is eventually treated in Chapter 5. Chapter 4 contains design that are useful for model-free approaches such as local regression and Chapter 6 describes design techniques that can be applied for making variogram estimation more efficient. The ultimative aim, also treated in Chapter 6, is to be able to construct designs with overall efficiency for spatial estimation.

Station number	$x_{i[1]}$	$x_{i[2]}$	$\bar{y}(x_i)$	N_T
411	-0.3503	0.1227	15.30	42
429	0.4418	-0.0620	59.31	40
849	0.3491	0.0114	58.26	37
854	0.2446	-0.0228	70.23	24
1502	0.4913	-0.0088	57.39	42
1584	-0.1262	0.0157	53.90	30
1591	-0.2115	0.1144	11.07	21
2046	-0.7700	0.2095	14.58	21
2047	-0.5384	0.0971	6.63	21
2048	-0.4275	0.1338	7.30	21
2049	-0.3183	0.0692	24.14	21
2051	-0.1744	-0.0216	29.37	21
2052	-0.1939	0.0276	36.92	21
2053	-0.1666	-0.1295	36.32	21
2054	-0.1622	-0.0507	38.72	21
2055	-0.1420	0.0811	45.83	20
2057	-0.0635	-0.0578	38.90	21
2058	-0.0470	0.0541	38.80	20
2059	-0.0456	-0.1025	23.99	21
2060	0.0339	0.0333	52.52	19
2061	0.1779	0.0048	63.59	24
2062	0.1957	-0.0675	66.52	19
2063	0.2958	-0.0026	52.24	21
2064	0.4482	-0.0276	66.52	21
2065	0.4340	0.0344	61.49	21
2066	0.6238	0.0521	64.69	19
2067	0.6700	0.0595	41.38	21
2070	-0.5590	0.1526	10.47	19
2071	-0.5806	0.2050	14.81	21
2072	0.0646	-0.0840	84.12	21
2128	-0.5222	0.0789	7.42	21
5319	-0.2010	0.0808	54.37	3
5320	0.0740	0.0003	116.00	3
5321	0.2463	-0.0885	119.00	1
5322	0.5702	0.0456	47.97	3
5323	0.2931	0.0279	55.57	3

TABLE 2.2. Monitoring station coordinates, averages and numbers of nonmissing observations.

References

Altman, N.S. (1990). Kernel smoothing of data with correlated errors. *Journal of the American Statistical Association*, 85(411):749–758.

Box, G.E.P. and Cox, D.R. (1964). An analysis of transformations. *Journal of the Royal Statistical Society, Series B*, 26:211–243.

Brockwell, P. and Davis, R.A. (1991). *Time Series. Theory and Methods*, 2nd edition. Springer—Verlag, Berlin.

Buja, A., Hastie, T. and Tibshirani, R. (1989). Linear smoothers and additive models with discussion. *The Annals of Statistics*, 17(2):453–555.

Cassel, C., Särndal, C. and Wretman, J.H. (1977). *Foundations of Inference in Survey Sampling*. Wiley, New York.

Chiles, J., and Delfiner, P. (1999). *Geostatistics. Modeling Spatial Uncertainty*. Wiley Series in Probability and Statistics, New York.

Cleveland, W.S. (1979). Robust locally weighted regression and smoothing scatterplots. *Journal of the American Statistical Association*, 74(368):829–836.

Cleveland, W.S., Grosse, E.H. and Shyu, M.J. (1992). Local regression models. In Chambers, J.M. and Hastie, T., editors, *Statistical Models in S*. Chapman and Hall, New York, 309–376.

Cleveland, W.S. and Loader, C. (1996). Smoothing by local regression: Principles and methods. In Härdle, W. and Schimek, M.G., editors, *Statistical Theory and Computational Aspects of Smoothing*. Physica Verlag, Heidelberg, 10–49.

Cressie, N. (1985). Fitting variogram models by weighted least squares. *Mathematical Geology*, 17:563–586.

Cressie, N. (1993). *Statistics for Spatial Data*, revised edition. John Wiley and Sons, New York.

del Pino, G. (1989). The unifying role of iterative generalized least squares in statistical algorithms. *Statistical Science*, 4:394–408.

Droge, B. (1996). Some comments on cross-validation. In Härdle, W. and Schimek, M.G., editors, *Statistical Theory and Computational Aspects of Smoothing*. Physica Verlag, Heidelberg, 178–199.

Dutter, R. (1985). *Geostatistik*. Teubner, Stuttgart.

Eubank, R.L. (1988). *Spline Smoothing and Nonparametric Regression.* Dekker, New York.

Fan, J. (1993). Local linear regression smoothers and their minimax efficiencies. *The Annals of Statistics,* 21(1):196–216.

Fedorov, V.V. (1972). *Theory of Optimal Experiments.* Academic Press, New York.

Fedorov, V.V. (1974). Regression problems with controllable variables subject to error. *Biometrika,* 61:49–56.

Fedorov, V.V. (1989). Kriging and other estimators of spatial field characteristics. *Atmospheric Environment,* 23(1):175–184.

Fedorov, V.V., Hackl, P., and Müller, W.G. (1993). Moving local regression: The weight function. *Journal of Nonparametric Statistics,* 2(4):355–368.

Fienberg, S.E. and Tanur, J.M. (1987). Experimental and sampling structures: Parallels diverging and meeting. *International Statistical Review,* 55:75–96.

Genton, M.G. (1998). Variogram fitting by generalized least squares using an explicit formula for the covariance structure. *Mathematical Geology,* 30(4):323–345.

Goldberger, A. (1962). Best linear unbiased prediction in the generalized linear regression model. *Journal of the American Statistical Association,* 57:369–375.

Gotway, C.A. (1990). Fitting semivariogram models by weighted least squares. *Computers & Geosciences,* 17:171–172.

Guttorp, P. and Sampson, P.D. (1994). Methods for estimating heterogeneous spatial covariance functions with environmental applications. In Patil, G.P., and Rao, C.R., editors, *Handbook of Statistics.* Elsevier Science, Amsterdam, 661–689.

Haas, T.C. (1990). Lognormal and moving-window methods of estimating acid deposition. *Journal of the American Statistical Association,* 85:950–963.

Haas, T.C. (1995). Local prediction of a spatio-temporal process with an application to wet sulfate deposition. *Journal of the American Statistical Association,* 90(432):1189–1199.

Haining, R. (1990). *Spatial Data Analysis in the Social and Environmental Sciences.* Cambridge University Press, Cambridge.

Härdle, W., Hall, P. and Marron, J.S. (1988). How far are automatically chosen regression smoothing parameters from their optimum? (with discussion). *Journal of the American Statistical Association*, 83:86–99.

Härdle, W. (1990). *Applied Nonparametric Regression*. Cambridge University Press, Cambridge.

Hastie, T.J. and Tibshirani, R.J. (1990). *Generalized Additive Models*. Chapman and Hall, London.

Hastie, T.J. and Tibshirani, R.J. (1993). Varying-coefficient models. *Journal of the Royal Statistical Society, Series B*, 55(4):757–796.

Host, G. (1999). Kriging by local polynomials. *Computational Statistics & Data Analysis*, 29:295–312.

Isaaks, E.H. and Srivastava, R.M. (1989). *An Introduction to Applied Geostatistics*. Oxford University Press, New York.

Journel, A.G. and Huijbregts, C.J. (1978). *Mining Geostatistics*. Academic Press, London.

Lamorey, G. and Jacobson, E. (1995). Estimation of semivariogram parameters and evaluation of the effects of data sparsity. *Mathematical Geology*, 27(3):327–358.

Maderbacher, M. and Müller, W.G. (1995). Comparing local fitting to other automatic smoothers. In Seeber G.U.H., Francis B.J., Hatzinger R., and Steckel-Berger G., editors, *Statistical Modelling*. Springer—Verlag, New York, 161–168.

Matheron, G. (1963). Principles of geostatistics. *Economic Geology*, 58:1246–1266.

McBratney, A.B. and Webster, R. (1986). Choosing functions for semivariograms of soil properties and fitting them to sampling estimates. *Journal of Soil Science*, 37:617–639.

McLain, D.H. (1971). Drawing contours from arbitrary data points. *The Computer Journal*, 17(4):318–324.

Müller, H.G. (1988). *Nonparametric Regression Analysis of Longitudinal Data*, volume 46 of Lecture Notes in Statistics. Springer—Verlag.

Müller, W.G. (1999). Least squares fitting from the variogram cloud. *Statistics & Probability Letters*, 43:93–98.

Munn, R.E. (1981). *The Design of Air Quality Monitoring Networks*. MacMillan, London.

Opsomer, J.D. (1996). Estimating an unknown function by local linear regression when the errors are correlated. *ASA Proceedings of the Statistical Computing Section*. Alexandria, VA, 102–107.

Pázman, A. (1993). *Nonlinear Statistical Models*. Mathematics and Its Applications. Kluwer Academic Publishers, Dordrecht.

Pelto, C.R., Elkins, T.A. and Boyd, H.A. (1968). Automatic contouring of irregularly spaced data. *Geophysics*, 33(3):424–430.

Pilz, J., Spoeck, G. and Schimek, M.G. (1997). Taking account of uncertainty in spatial covariance estimation. In Baafi, E. and Schoefield, N., editors, *Geostatistics Wollongong '96*. Kluwer Academic Publishers, Amsterdam, 1:302–313.

Rajagopalan, B. and Lall, U. (1998). Locally weighted polynomial estimation of spatial precipitation. *Journal of Geographic Information and Decision Analysis*, 2(2):48–57.

Ripley, B.D. (1981). *Spatial Statistics*. Wiley, New York.

Roddenberry, G. (1964). Star Trek.

Ruppert, D., Wand, M.P., Holst, U. and Hössjer, O. (1997). Local polynomial variance-function estimation. *Technometrics*, 39(3):262–273.

Sampson, P. and Guttorp, P. (1992). Nonparametric estimation of nonstationary covariance structure. *Journal of the American Statistical Association*, 87:108–119.

Seber, G.A.F. and Wild, C.J. (1989). *Nonlinear Regression*. Wiley, New York.

Stein, M.L. (1988). Asymptotically efficient prediction of a random field with misspecified covariance function. *The Annals of Statistics*, 16(1):55–63.

Stone, M. (1974). Cross-validatory choice and assessment of statistical predictions. *Journal of the Royal Statistical Society, Series B*, 36:111–147.

Thiebaux, H.J. and Pedder, M.A. (1987). *Spatial Objective Analysis, with Applications in Atmospheric Science*. Academic Press, London.

Wackernagel, H. (1995). *Multivariate Geostatistics*. Springer—Verlag, Berlin—Heidelberg.

Zhang, X.F., van Eijkeren, J.C.H. and Heemink, A.W. (1995). On the weighted least-squares method for fitting a semivariogram model. *Computers & Geosciences*, 21(4):605–606.

Zimmerman, D.L. and Cressie, N. (1992). Mean squared prediction error in the spatial linear model with estimated covariance parameters. *Annals of the Institute of Statistical Mathematics*, 44(1):27–43.

Zimmerman, D.L. and Zimmerman, M.B. (1991). A comparison of spatial semivariogram estimators and corresponding ordinary kriging predictors. *Technometrics*, 33(1):77–91.

3
Fundamentals of Experimental Design

"A motorcycle mechanic, [...], who honks the horn to see if the battery really works is informally conducting a true scientific experiment. [...] To design an experiment properly he has to think very rigidly in terms of what directly causes what."
Pirsig (1974)

Early statistical investigations about how to plan experiments efficiently originated almost entirely from agricultural problems, see the well-known pioneering monograph by Fisher (1935). Therein the emphasis is laid on detecting the influences of discrete factors leading to heavy use of combinatorial principles in the construction of the experimental designs (e.g. Latin squares, Balanced Incomplete Block Designs etc.). A whole branch of theory for *categorical designs* developed along these lines and one can find a vast number of publications devoted to it (one of the most widely used textbooks is Cox (1958)). However, such designs are of limited use in the spatial setting, where the essential factors, the coordinates, are continuous rather than discrete (they may be still useful for so-called neighboring problems, see Martin (1996) for a recent review).

For classical regression relationships Box and Wilson (1951) suggested the use of the in what followed so-called *response surface de-*

signs. Therewith they initiated a school that proceeds to produce many results that are useful for spatial design theory (see Myers and Montgomery (1995) for a recent monograph). Their aim is to develop standard designs (such as fractional factorials or central composite designs) for regularly shaped design regions in order to comply with a list of requirements (see Box and Draper (1975)) for proper experimentation.

A different approach is adopted by the third large branch of experimental design theory, the so-called *optimum design of experiments* (*o.d.e.*, predominantly put forward by Kiefer (1985)). Here, the main construction principle for a well designed experimental plan is as follows. Asserting hypothesis-testing or parameter estimation as the aim of the experiment, the task remains to take a (possibly given) number of measurements in a way, that either the power of the test or the precision of the estimator is maximized. Additional requirements such as the unbiasedness of the statistical method, restrictions on the experimental conditions (e.g. to a certain experimental region), etc. usually enter as side-conditions of the maximization problem. The advantage of the *o.d.e.* approach is that it allows the application of computational algorithms for the construction of tailor-made designs. This is of particular importance in the spatial setting where the experimental region hardly ever is of regular shape and other experimental conditions frequently violate symmetry properties.

Moreover, often do the designs obtained by running *o.d.e.* algorithms correspond very closely to those derived from combinatorial principles or practicability rules such as given in Box and Draper (1987). Therefore, in what follows it will mainly be referred to results from *o.d.e.* theory, but the many parallels to the other two branches of design theory will be pointed out.

Since the pioneering monograph by Fedorov (1972) there have appeared a number of book-length treatments on *o.d.e.* stressing various aspects of the theory and its applications, such as Bandemer *et al.* (1977) (encyclopedic), Silvey (1980) (concise and dense), Pázman (1986) (algebraic), and Atkinson and Donev (1992) (application-oriented). Two purely theoretical surveys of the current state of knowledge can be found in Pukelsheim (1993) and Fedorov and Hackl (1997).

For a more formal introduction, let us call the set

$$\xi_N = \left\{ \begin{array}{cccc} p_1, & p_2, & \ldots, & p_n \\ x_1, & x_2, & \ldots, & x_n \end{array} \right\}$$

a (normalized) exact design of a corresponding experiment

$$\left\{ \begin{array}{cccc} y_{11}, y_{12}, \ldots, y_{1N_1}; & y_{21}, y_{22}, \ldots, y_{2N_2}; & ; \ldots; & y_{n1}, y_{n2}, \ldots, y_{nN_n} \\ x_1; & x_2; & ; \ldots; & x_n \end{array} \right\},$$

of $N = \sum_{i=1}^{n} N_i$ observations. Here the y_{ij} denote the observed values at the design points (coordinates) $x_i \in \mathcal{X}$. Note that we allow for replicate measurements at a specific design point, so that the weights $p_i = N_i/N$ (accordingly with $\sum_{i=1}^{n} p_i = 1$) are proportional to the numbers of replications. The weights p_i may also be regarded as precision or duration of the measurements. Thus the 'pure' problem in o.d.e. can be formulated as finding the optimum design

$$\xi_N^* = \arg \max_{\xi_N \in \Xi_N} \mathcal{J}(\xi_N), \qquad (3.1)$$

where $\mathcal{J}(\xi_N)$ denotes some measure of information obtained from an experiment based on the design ξ_N, and Ξ_N the set of all admissible exact N-observation designs.

If the number of observations is large enough, one can use a continuous design ξ and a respective formulation of (3.1) instead of the discrete ξ_N and proceed with the more comfortable tools of continuous mathematics. This was firstly proposed in the pioneering paper by Kiefer (1959). We characterize a continuous (normalized) design ξ through a discrete probability measure $\xi(x)$ with

$$\sum_{x \in \mathcal{X}} \xi(x) = 1 \qquad \left\{ \begin{array}{ll} \xi(x) > 0, & x \in S_\xi \\ \xi(x) = 0, & x \in \mathcal{X} \setminus S_\xi \end{array} \right.,$$

where $S_\xi \subset \mathcal{X}$ stands for a set of support with a finite number of points.

Such designs are called approximative designs, in contrast to exact discrete designs applied in practice, since frequently only after rounding one gets a feasible experimental plan. In this chapter concentration is on the approximate theory, suggestions on rounding in order to get exact designs can be found in Pukelsheim (1993). However, in later chapters methods that directly produce exact designs such that rounding is not necessary will be presented.

3.1 Information Matrices

How can the information $\mathcal{J}(\xi)$ due to a design ξ be formally described in a real experiment? If one wants to apply methods from o.d.e. theory, he or she has initially to pose the question, which mathematical model serves as a sufficiently good description of the observed field.

In Chapter 2 the basic model in spatial statistics, the random field (2.1) with covariance function (2.2), was presented. Our primary focus will be to estimate its parameters β or linear functionals $l^T\beta$ of them. In this chapter the simplified (uncorrelated) version of (2.2), i.e.

$$c(x, x'; \theta) = \delta_{x,x'}\sigma^2(x, \theta),$$

is considered. Design aspects of the more general case are treated in Chapter 5.

We will also firstly assume that $\sigma^2(\cdot, \cdot)$ is completely known up to a multiplicative factor, such that by a proper rescaling we arrive at $\sigma^2(\cdot, \cdot) \equiv 1$ without loss of generality. In the vicinity of the true β a linearization of $\eta(\cdot, \cdot)$ may be sufficiently exact and thus (almost) all the information about β (at least with respect to estimating linear functionals of it, see Pázman (1986)) from the experiment is contained in its so-called (standardized or average) information matrix

$$M(\xi, \beta) = \sum_{x \in \mathcal{X}} \dot{\eta}(x, \beta)\dot{\eta}^T(x, \beta)\xi(x). \tag{3.2}$$

It is additive on ξ, symmetric and positive definite (if the set of support S_ξ is extended enough). A thorough discussion on the situation for singular $M(\cdot, \cdot)$ is given in Pukelsheim (1993). Moreover, the set of all information matrices $\mathcal{M} = \{M(\xi, \beta) : \xi \in \Xi\}$, where Ξ is the set of all approximate designs on $\mathcal{X}$, is convex and compact (cf. Silvey (1980)). Also any matrix $M(\cdot, \cdot) \in \mathcal{M}$ can be constructed as a convex linear combination of no more than $\frac{m(m+1)}{2} + 1$ elements and thus we can limit the search for an optimum design to the set $\bar{\Xi}$, which contains all designs with no more than $\frac{m(m+1)}{2} + 1$ support points (cf. e.g. Fedorov (1972)).

Note that the inverse of the (unstandardized) information matrix corresponding to (3.2) coincides with the asymptotic dispersion matrix of the estimator (2.3), so that we have

$$\text{Cov}[\hat{\beta}] \simeq \frac{1}{N}M^{-1}(\xi, \hat{\beta}),$$

under rather mild assumptions about the consistency of $\hat{\beta}$ (and the curvature of $\eta(\cdot, \cdot)$ respectively). For the improvement of the approximations in this context Pázman and Pronzato (1992) suggest to exploit distributional information about the errors and a correspondingly modified setup. An interesting account of various interpretations of the information matrix in a deterministic setting is given in Fedorov and Hackl (1997).

Unfortunately $M(\cdot,\cdot)$ shows two features that prohibit its direct use as a measure of information $\mathcal{J}(\xi)$ in a continuous version of (3.1). Firstly, the ordering of the designs, which is induced by the Loewner ordering of the corresponding information matrices, i.e. a design is ξ' uniformly worse than a design ξ iff

$$M(\xi,\beta) - M(\xi',\beta) \text{ is positive definite,}$$

is generally only a partial ordering on the set of all designs Ξ. Thus for a complete comparison of designs we will have to introduce a suitable scalar function $\Phi[M(\xi,\beta)]$ to arrive at a complete ordering of candidate designs. Necessary properties for $\Phi[\cdot]$ and various popular choices are presented in the next section.

Locally Optimum Designs

Secondly, in general $M(\xi,\beta)$ depends upon the true values of the parameter and not only upon the design ξ, which evidently leads to a circular problem: for finding a design that estimates β efficiently it is required to know its value in advance. The most commonly used approach proposed by Chernoff (1953) is to evaluate the linearization of $\eta(\cdot,\cdot)$ at a best guess β_0 and use $\dot{\eta}(x,\beta_0)$ in (3.2) to find an approximate information matrix $M(\xi,\beta_0)$. This is valid only locally and thus designs resulting from procedures based upon it are called locally optimum designs.

The original optimization problem (3.1) is now transformed to

$$\xi^* = \arg\max_{\xi \in \Xi} \Phi[M(\xi,\beta_0)]. \tag{3.3}$$

This (in sometimes modified form) will be the key problem considered in this monograph. Note that the original setup had to undergo several approximations to arrive at (3.3) and thus the solution will be generally not the same. However, we can still expect to come sufficiently close and therewith obtain reasonably efficient designs for practical purposes. In order to simplify notations, from the next section on the argument β_0 will be omitted when it leads to no ambiguity.

There are other approaches for treating the problem of the dependence of $M(\cdot,\cdot)$ upon the true values of β. Cook and Fedorov (1995) suggest to add to (3.3) the side condition

$$\text{s.t. } \Phi[M(\xi,\beta)] \geq \kappa \geq 0, \qquad \forall \beta \in \Omega,$$

to ensure a minimum level of information even in the case that the initial guess is wrong. An algorithm for its solution is proposed by Müller (1995).

Bayesians like to use an average

$$E_\beta\{\Phi[M(\xi,\beta)]\} = \int \Phi[M(\xi,\beta)]\pi(\beta)d\beta, \tag{3.4}$$

where $\pi(\beta)$ is a suitable prior distribution on β, rather than a local guess (cf. Chaloner and Verdinelli (1995)). A similar, though prior-free approach is adopted by Firth and Hinde (1997).

If observations can be taken sequentially the maximum amount of information of an additional observation can be gained by using an adaptive procedure:

$$x_{i+1}^\# = \arg\max_{x \in \mathcal{X}} \Phi[M(x, \hat{\beta}_{\{i\}})],$$

where $\hat{\beta}_{\{i\}}$ is the estimate based upon the first i observations. The sequence $\xi_N^\# = \{x_1^\#, \ldots, x_N^\#\}$ then forms what is known as a fully sequential design (for properties see e.g. Ford and Silvey (1980)). It is remarkable that only in this approach the actual observations from the experiment do have influence on the structure of the resulting design. For details about this and other properties of sequential experimentation and their decision-theoretic impact refer to the monograph by Chernoff (1972).

A combination of the two approaches is suggested by Chaloner (1993), who with the help of a simple model devise a batch sequential Bayesian method, which unifies theoretical (use of the best available information about β) and practical (information gain through new data) advantages. Related investigations on the asymptotic behavior of similar design strategies for the same model can be found in Müller and Pötscher (1992).

Linear Regression

Note that in one of the most useful cases this second problem, the dependence of the information matrix upon the true values of the parameters, does not arise. If $\eta(x,\beta)$ is linear in β (or can be respectively transformed) $\dot{\eta}(\cdot)$ does not depend upon β and hence the same applies to $M(\xi,\cdot)$. As an example consider the two-dimensional second degree polynomial, which usually is a reasonable approximation of a unimodal surface and is used as the basic model in response surface methodology (cf. Box et al. (1978)):

$$\eta(x,\beta) = \beta_0 + \beta_1 x_{[1]} + \beta_2 x_{[2]} + \beta_{11} x_{[1]}^2 + \beta_{22} x_{[2]}^2 + \beta_{12} x_{[1]} x_{[2]}. \quad (3.5)$$

Here, simply $\dot{\eta}^T(x) = (1, x_{[1]}, x_{[2]}, x_{[1]}^2, x_{[2]}^2, x_{[1]} x_{[2]})$ (indices in brackets denote coordinates rather than running point numbers). An assignment of a best guess (or any other methods from above) is therefore not necessary and one can proceed with direct maximization of $\Phi[\cdot]$.

The linear case is treated extensively in all books on optimum design. Special attention towards the nonlinear setup is devoted by the survey paper of Ford et al. (1989). Note that the design problem in this setup is similar to the one in the case of limited dependent variables modeled by generalized linear models as emphasized by Atkinson and Haines (1996). Designs that are useful if a linear (or linearized) part is contaminated by a nonlinear term are presented in Wiens (1992).

Distributed Parameter Systems

When one is faced with spatial phenomena including a temporal dimension (in the form of so-called distributed parameter systems), the dynamical behavior of the response is often modeled by partial differential equations (e.g. of advection diffusion type in pollution problems), i.e.

$$\frac{\partial \eta(x,t)}{\partial t} = f\left(x, t, \eta, \frac{\eta}{\partial x_{[1]}}, \frac{\eta}{\partial x_{[2]}}, \frac{\eta^2}{\partial x_{[1]}^2}, \frac{\eta^2}{\partial x_{[2]}^2}, \beta\right), \quad t \in (0,T), \quad (3.6)$$

with appropriate initial and boundary conditions.

The corresponding information matrix given by

$$M(\xi, \beta; T) = \sum_{x \in \mathcal{X}} \xi(x) \int_0^T \dot{\eta}(x, t; \beta) \dot{\eta}^T(x, t; \beta) dt$$

is used in place of (3.2) (see Quereshi et al. (1980)). With mild adaptations the numerical design optimization methods given in this chapter can still be used. For an excellent survey on design for distributed parameter systems and further details refer to Uciński (1999).

3.2 Design Criteria

Useful Properties of Design Criteria

For a function $\Phi[\cdot]$ to qualify as a reasonable design criterion it has to satisfy several properties, partly for practical, partly for theoretical reasons. First, the partial ordering induced by the information matrix should be maintained by the complete ordering due to the criterion, i.e. if $M(\xi) - M(\xi')$ is positive definite then $\Phi[M(\xi)] > \Phi[M(\xi')]$.

Second, it is useful for $\Phi[M(\xi)]$ to be a concave function, i.e.

$$\Phi[(1-\alpha)M(\xi')+\alpha M(\xi'')] \geq (1-\alpha)\Phi[M(\xi')]+\alpha\Phi[M(\xi'')], \quad 0 \leq \alpha \leq 1,$$

and there has to be at least one design for which $\Phi[M(\xi)]$ attains a nonzero value. Furthermore the directional derivative of $\Phi[\cdot]$ with respect to the design measure in the direction of a design ξ' has to be linear in ξ' and for a one-point support $S_{\xi'} = \{x\}$ it is of the form

$$\lim_{\alpha \to 0} \frac{\Phi[(1-\alpha)M(\xi) + \alpha M(\xi')] - \Phi[M(\xi)]}{\alpha} = \phi(x,\xi) - \Phi'[M(\xi)],$$

where $\phi(x,\xi)$ is sometimes (cf. Fedorov and Hackl (1997)) called the sensitivity function for reasons given in the next sections, and $\Phi'[M(\xi)] \equiv \text{tr}\{M(\xi)\nabla\Phi[M(\xi)]\} = -\Phi[M(\xi)]$ for many criteria. Here $\nabla\Phi[M]$ denotes the gradient matrix of $\Phi[M]$ with entries $\frac{\partial \Phi[M]}{\partial M_{ii'}}$; $i, i' = 1, \ldots, m$.

Criteria Satisfying Those Properties

In the case of Gaussian errors one can depict the confidence regions for the estimator (2.3) as ellipsoids in $\mathbb{R}^m$. More generally, the confidence ellipsoids for the estimators of some linear functionals $L^T\beta$ are given by

$$(\hat{\beta} - \beta)^T L[L^T M^{-1}(\xi) L]^{-1} L^T (\hat{\beta} - \beta) = \kappa,$$

where $L = (l_1, \ldots, l_k)$. The size (and form) of these ellipsoids depend upon the design ξ and their minimization is very closely related to (3.3) for some specific functionals $\Phi[\cdot]$. Note that L may depend upon the guess β_0 as well (as is e.g. the case in Atkinson et al. (1993) and Fedorov et al. (1995)). For a higher order approximation of the confidence regions and a resulting design criterion see Hamilton and Watts (1985).

3.2 Design Criteria

For illustrative purposes let us assume that the response function is a special form of (2.14), namely a Scheffé polynomial (a plane with an interaction term)

$$\eta(x,\beta) = \beta_0 + \beta_1 x_{[1]} + \beta_2 x_{[2]} + \beta_3 x_{[1]} x_{[2]},$$

and that we intend to estimate the functionals $\beta_0 + \frac{1}{2}(\beta_1+\beta_2) + \frac{1}{4}\beta_3$ and β_0, which correspond to the averages of this surface over $[0,1]^2$, and $[-1,1]^2$, respectively. This yields $l_1 = (1,0,0,0)$ and $l_2 = (1,\frac{1}{2},\frac{1}{2},\frac{1}{4})$ and its confidence ellipsoids for three different designs are given in Figure 3.1.

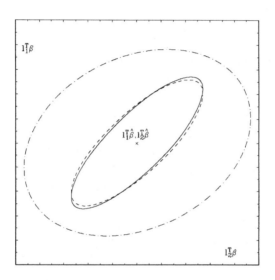

FIGURE 3.1. Confidence ellipsoids corresponding to different designs

The ellipsoid with the solid line is the one with the smallest possible volume and its corresponding design ξ_D^* is called (generalized) **D − optimum**. Here

$$\Phi[M(\xi)] = -\ln|L^T M^-(\xi) L|,$$

where the superscript $-$ denotes the Moore-Penrose g-inverse to allow for singular designs. The simple version where $L = I_m$ (i.e. the minimization of the generalized variance of $\hat{\beta}$) is the most popular choice in experimental design (in the nonlinear context it was introduced by Box and Lucas (1959)). It has several additional advantageous properties (e.g. scale-invariance) and will be extensively applied in this monograph. Cases where L is chosen to contain only 0 or 1, such that emphasis is on particular subsets of parameters are often referred to as D_s-optimality (see e.g. Atkinson and Donev (1992)). Here, a similar notation will be adopted in referring to the more general case as D_L-optimality. In the above setting the D_L-optimum design has support on the vertices of the experimental region $\mathcal{X} = [-1,1]^2$ with corresponding measures $\xi^*_{D_L}(-1,-1) = 0.230$, $\xi^*_{D_L}(-1,1) = \xi^*_{D_L}(1,-1) = 0.164$, and $\xi^*_{D_L}(1,1) = 0.442$. The notation here corresponds to $\xi(x_i) = p_i$.

The design with measures $\xi^*_{A_L}(-1,-1) = 0.187$, $\xi^*_{A_L}(-1,1) = \xi^*_{A_L}(1,-1) = 0.216$, and $\xi^*_{A_L}(1,1) = 0.381$ yields the dashed ellipsoid and is calculated from maximizing

$$\Phi[M(\xi)] = \mathrm{tr}[L^T M^-(\xi) L]^{-1}.$$

It is called the (generalized) **A-optimum** (A_L-optimum) design and corresponds geometrically to minimizing the distance from the center of the ellipsoid to the corners of the enclosing rectangle. The dash-dot ellipsoid in Figure 3.1 represents an arbitrary design for comparison purposes.

The D-optimum design for the whole parameter set (i.e. $L = I_m$) is also supported on the four vertices of the design region. This can be easily seen from Figure 3.2, which represents the so-called induced design space, the set generated by $\mathcal{F} = \{\dot{\eta}(x) : x \in \mathcal{X}\}$. For D-optimum designs the support points must coincide with the intersection points of this set with a minimum volume ellipsoid centered at the origin, which contains $\mathcal{F}$ (cf. Silvey (1980) or Haines (1993)). Moreover, in this case the exact D-optimum 4-point design coincides with the approximate design, since the exact m-point optimum design always forms the maximum volume simplex by its support points, as is evidently the case here. Thus we have $\xi^*_D(-1,-1) = \xi^*_D(-1,1) = \xi^*_D(1,-1) = \xi^*_D(1,1) = 0.25$. Note that this is the same optimum design as for the purely linear response, the plane without interaction term.

A simpler method for determining whether a given design is the approximate optimum design is provided by the fact that (when a criterion satisfies the above properties)

$$\phi(x,\xi^*) \leq \Phi'[M(\xi^*)] \qquad \forall x \in \mathcal{X}, \tag{3.7}$$

see e.g. Fedorov and Hackl (1997). In the case of D-optimality the derivative is

$$\phi(x,\xi^*_D) = \dot{\eta}(x)^T M^{-1}(\xi^*_D) \dot{\eta}(x), \tag{3.8}$$

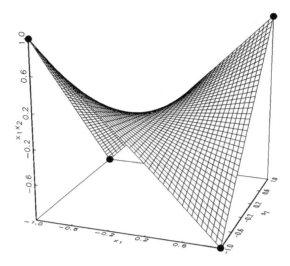

FIGURE 3.2. Induced design space $\mathcal{F}$ and intersection points.

which is depicted in Figure 3.3. The function $\phi(x, \xi_D^*)$ takes its maximal value $\Phi'[M(\xi_D^*)] = m = 4$ just at the vertices and thus (3.7) is fulfilled for ξ_D^*. From this figure it can also be seen that the iso-lines do not form circles as in the case of a purely linear response and the optimum design on a region other than the square can be different for the two models. Specifically, for the unit-circle the D-optimum 4 point design for the model with interaction term is only attained at the points, where the diagonals of the square intersect the circle, whereas for the model without interaction the vertices of any square inscribed into the circle are D-optimum points.

When we are interested in just a single linear function (i.e. $k = 1$) the above design criteria coincide to so-called **c-optimality**, where

$$\Phi[M(\xi)] = \mathrm{var}^{-1}(l^T \beta).$$

In this case a strong geometrical tool to determine the optimum design is to find the intersection point of the vector l with the boundary of the so-called Elfving set (Elfving (1952)), which is the convex hull of $\mathcal{F}$ and its reflection with respect to the origin. In the case $l = l_2$ from above the vector $l = (1, 0.5, 0.5, 0.25)$ intersects directly with $\mathcal{F}$ at $x_c^* = (0.5, 0.5)$, which is thus the (singular) c-optimum design (cf. Ford et al. (1992)). This is not surprising, since x_c^* is just the central point of the region $[0, 1]^2$ for which an average estimate was sought for.

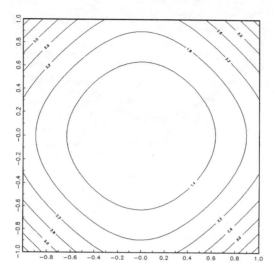

FIGURE 3.3. Contour map of the sensitivity function on a unit square.

That it is not necessarily so that the optimum exact design can be found on the support points of the optimum approximate design can be demonstrated by solving the problem of estimation of the average response over $[-1,1]^2$ and $[0,1]^2$ in the more general model (3.5). Here, $l_1 = (1,0,0,0,0,0)$ and $l_2 = (1,\frac{1}{2},\frac{1}{2},\frac{1}{3},\frac{1}{3},\frac{1}{4})$ and the D-optimum approximate design $\xi_{D_L}^*$ is a (singular) 5 point design, with measures $\xi_{D_L}^*(-1,-1) = 0.096, \xi_{D_L}^*(-1,1) = \xi_{D_L}^*(1,-1) = 0.019, \xi_{D_L}^*(1,1) = 0.366$, and $\xi_{D_L}^*(0,0) = 0.500$, the respective exact 5 point design, however, is $\xi_{5,D_L}^* = \{(-0.4,-0.4),(0,0),(0.3,0.3),(0.4,0.4),(1,1)\}$. Figure 3.4 gives the confidence ellipsoids corresponding to the designs $\xi_{D_L}^*$ (solid), ξ_{5,D_L}^* (dashed) and the exact 5 point design on the support of $\xi_{D_L}^*$ (closely dashed).

Other Popular Optimality Criteria

If there is no focus on a special set of linear combinations $L^T\beta$ to be estimated, but it is desired to estimate all possible of them reasonably well, one could aim for

$$\Phi[M(\xi)] = -\max_{l} \frac{l^T M^-(\xi) l}{l^T l},$$

i.e. minimization of the variance of the worst estimated contrast. This is geometrically equivalent to minimizing the longest halfaxis of the

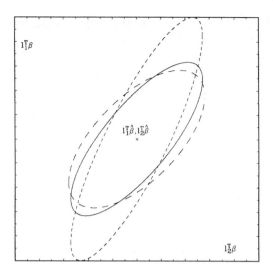

FIGURE 3.4. Confidence ellipsoids for various designs.

confidence ellipsoid. The criterion is called **E-optimality** and is rarely used due to its unfavorable theoretical properties (cf. Dette and Haines (1994)).

Another straightforward criterion with similar limitations leads to **maximum variance optimality**

$$\Phi[M(\xi)] = -\max_i M_{ii}^-(\xi).$$

The quality of experimental designs cannot only be judged by good estimation of parameters, but also via precision of predictors of the response values. This leads to a different type of natural criteria. Let the variance of a predictor $\hat{y}(x_0)$ be denoted by $d[x_0, M(\xi)] = \dot{\eta}^T(x_0) M^{-1}(\xi) \dot{\eta}(x_0)$, then one calls

$$\Phi[M(\xi)] = -\max_{x \in \mathcal{Z}} w(x) d[x, M(\xi)]$$

the (generalized) **G-optimality** criterion. Here, the weighting function $w(x)$ reflects differences in the importance of precise predictions over a region of interest $\mathcal{Z}$. A criterion that is based upon a similar principle leads to α-**optimality**, where

$$\Phi[M(\xi)] = \int_{\mathcal{Z}} w(x) d^{-1}[x, M(\xi)] dx.$$

Surveys on the use of the various optimality criteria give e.g. Hedayat (1981) or Shah and Sinha (1989). Note that the presented criteria are useful mainly in the independent error case, whereas for the general setting different concepts (e.g. criteria based upon the entropy) may be required. Some of these and their relationship to the classical criteria from this section will be presented in Chapters 4 and 5.

Relationships among Criteria

A first connection between the above optimality criteria is revealed if one regards the eigenvalues $\lambda_1, \ldots, \lambda_m$ of $M^{-1}(\xi)$. The D-criterion aims at minimizing their product, the A-criterion their sum, and the E-criterion their minimum. They can thus be, as proposed by Kiefer (1975), summarized as members of a wider class of criteria

$$\Phi_p[M(\xi)] = \left(\frac{1}{m}\sum_{i=1}^m \lambda_i^p\right)^{-\frac{1}{p}},$$

the so-called Φ_p-**optimality** criterion (where D-optimality corresponds to $p \to 0$, A-optimality to $p = 1$ and E-optimality to $p \to \infty$). This leads to the proposal of various authors (see e.g. Dette et al. (1995) or Mirnazari and Welch (1994)) for finding robust designs in this or a wider class of criteria. The concept of Φ_p-optimality was successfully applied to unify random and deterministic sampling by Cheng and Li (1987). The formulation of design criteria based directly upon the eigenvalues λ_i has also found considerable appeal in applications, for examples and further references see Curtis (1999).

One of the most important consequences of approximate theory is the possibility of identifying further relations between different optimality criteria. The main result, the so-called **general equivalence theorem**, is a generalization of the fundamental theorem of Kiefer and Wolfowitz (1960). It was formulated for the nonlinear case by Whittle (1973) and White (1973), and was given in its general form by Kiefer (1974), Fedorov (1980), and Silvey (1980). It states that the problems yielding the optimum design

$$\xi^* = \arg\max_\xi \Phi[M(\xi)] \equiv \arg\min_\xi \max_{x\in\mathcal{X}} \phi(x,\xi) \tag{3.9}$$

are equivalent and that moreover

$$\phi(x,\xi^*) = \Phi'[M(\xi^*)] \text{ for all } x \in S_{\xi^*}. \tag{3.10}$$

Observing that for D-optimality $\phi(x,\xi) = d[x, M(\xi)]$, cf. (5.22), we arrive at the equivalence of D- and G-optimality (for $\mathcal{Z} = \mathcal{X}$) and hence at the celebrated equivalence theorem of Kiefer and Wolfowitz (1960).

3.3 Numerical Algorithms

The sensitivity function $\phi(\cdot,\cdot)$ completely determines the support of any optimum design. It 'senses' the influence of a particular point $x^{(0)}$ on the design measure; $\phi(x^{(0)}, \xi) > \Phi'[M(\xi)]$ indicates that additional experimental effort at this location will pay off with respect to an increase in the criterion, cf. (3.7). It is therefore natural to suggest in a particular situation (with an existing design $\xi_{(0)}$) to construct a new design $\xi_{(1)}$ by increasing the measure at the point $x^{(0)}$ and reducing it everywhere else, i.e.

$$\xi_{(1)}(x) = (1 - \alpha^{(0)})\xi_{(0)}(x) + \alpha^{(0)}\delta_{x,x^{(0)}} \qquad \forall x \in \mathcal{X}, \qquad (3.11)$$

where $0 < \alpha^{(0)} < 1$. This can be done iteratively ($s = 0, 1, 2, \ldots$) in the hope that eventually no point $x^{(s)}$ with

$$\phi(x^{(s)}, \xi_{(s)}) > \Phi'[M(\xi_{(s)})]$$

remains, which indicates, cf. (3.10), that $\xi_{(s)}$ has converged to the optimum design ξ^*. Algorithms based upon this idea of one-point correction are predominant in the literature; prototypes have been proposed by Wynn (1970) and Fedorov (1972).

The most common choice for the determination of the candidate point for increasing the measure is

$$x^{(s)} = \arg\max_x \phi(x, \xi_{(s)}),$$

although it would be sufficient to choose any $x^{(s)}$ for which the sensitivity function is large.

Convergence of these type of algorithms is ensured in most the cases (see e.g. Atwood (1973)) if $\alpha^{(s)}$ is chosen as a properly diminishing sequence (with $\lim_{s\to\infty} \alpha^{(s)} = 0$), preferably

$$\alpha^{(s)} = \arg\max_\alpha \Phi[M(\xi_{(s+1)})],$$

which is known as the steepest ascent rule and may be difficult to be evaluated. It is usually sufficient to find a sequence, e.g. $\alpha^{(s)} = \frac{1}{N+s}$, such that, on the one hand, the search is assured not to stop too early ($\sum_{s=0}^{\infty} \alpha^{(s)} = \infty$) and on the other not to overshoot the target ($\sum_{s=0}^{\infty} (\alpha^{(s)})^2 < \infty$).

There are a number of straightforward modifications to improve the procedure, like decreasing the measure at points $x_-^{(s)}$ where $\phi(x_-^{(s)}, \xi_{(s)})$ is small in intermediate steps (backward excursion), with a respective negative weight

$$\alpha_-^{(s)} = \begin{cases} -\alpha^{(s)} & p(x_-^{(s)}) \geq \alpha^{(s)} \\ -\dfrac{p(x_-^{(s+1)})}{1-p(x_-^{(s+1)})} & p(x_-^{(s)}) < \alpha^{(s)} \end{cases},$$

or by merging support points that lie close to each other. For a detailed account of these refinements see Nguyen and Miller (1992). Molchanov and Zuyev (2000) have motivated their specific version of a backward-forward algorithm by not applying the commonly used directional derivative $\phi(\cdot, \cdot)$ for their calculations but rather the Gateaux derivative

$$\lim_{\alpha \to 0} \frac{\Phi[M(\xi) + \alpha M(\xi')] - \Phi[M(\xi)]}{\alpha},$$

thus employing the 'true' steepest ascent direction.

In principle, there is no reason to restrict the corrections to one-point measures. At every step of the algorithm one could identify a set of candidate points (possibly by evaluating higher order approximations of the directional derivative than $\phi(\cdot, \cdot)$) and modify the design measure accordingly (see e.g. Gaffke and Heiligers (1995)). A similar idea is the basis of the simultaneous correction algorithm proposed by Torsney (1977), where

$$\xi_{(s+1)}(x) = \frac{\xi_{(s)}(x)\phi^\lambda(x, \xi_{(s)})}{\sum_{x \in S_{\xi_{(s)}}} \xi_{(s)}(x)\phi^\lambda(x, \xi_{(s)})}, \qquad \forall x \in \mathcal{X}$$

and the free parameter λ is attempted to be chosen in an optimal way (cf. also Silvey et al. (1978)). Here, the initial design $\xi_{(0)}$ needs to be supported on the whole region $\mathcal{X}$, whereas for other algorithms it is usually sufficient to be nonsingular.

Note that both types of algorithms are only useful for calculating approximate designs. Adaptations for finding exact designs are sometimes straightforward, the simplest is due to Mitchell (1974), who proposes $\alpha^{(s)} = \frac{1}{N}$ and subsequent backward and forward excursions. A modified version of it can be used to augment points to (finite number of subsequent forward excursions) and delete points from (backward excursions) existing designs. Reviews of this type of algorithms are given in Cook and Nachtsheim (1980) and Atkinson and Donev (1992).

In the spatial setting, however, the number of potential sites is usually quite large and the set of candidate points is continuous so that we can expect that some rounding procedures (see e.g. Pukelsheim and Rieder (1992)) of the approximate designs calculated by the above algorithms will yield sufficiently good exact designs. Algorithms that directly produce exact designs and are particularly adapted to the special features of the spatial setting are presented in Chapter 5.

Algorithms that are based on quite different concepts have also been successfully applied for optimizing designs, such as simulated annealing (Haines (1987)), convex programming techniques (Gaffke and Mathar (1992)), or bundle trust methods (Wilhelm (1997)) to name just a few.

3.4 Further Design Topics Useful in the Spatial Setting

In spatial problems it is frequently the case that the design region $\mathcal{X}$ is not of a regular shape due to natural or political boundaries of the investigated area. Therefore, a simplification that might be useful is that of restricting the set from which each site in the design is chosen to a finite subset $\bar{\mathcal{X}}$ of $\mathcal{X}$. In some cases this restriction may come about naturally, as only a few sites in $\mathcal{X}$ are suitable for the kind of measurements being made. A classical example of this is meteorological station data.

In other situations, such as many soil studies, there is in principle no reason why attention must be restricted to a finite subset. In these cases, however, it simplifies matters greatly to maximize the design criterion by a grid search over a discretization $\bar{\mathcal{X}}$. Provided that the grid is not too fine, this procedure is computationally feasible; it is also more reliable than more sophisticated optimization algorithms (e.g, Nelder-Mead simplex) since the sensitivity function generally is not unimodal.

For computationally demanding setups (such as those presented in Chapter 5) it may be sufficient to compare optimum designs on random subsets of $\bar{\mathcal{X}}$ and then to propose the best of those as a reasonable approximation to the overall optimum design.

In situations where replicate measurements at sites are not possible, such as when making the measurement exhausts the experimental material, the set of candidate sites will get smaller as new sites are added to the design. To allow for this possibility, $\mathcal{X}_s$ shall denote the set of sites that can be added at the s-th stage of a sequential procedure.

Designs for Localizing Extremal Values

Frequently the aim in spatial data analysis is not only to estimate the parameters that determine a random field (or its averages over a specified region) but to detect at which point an extremal observation is likely to occur. This is particularly so in the search for mineral resources. In principle we can again utilize the generalized design criteria given Section 3.2, when L is determined from $\frac{\partial \eta(x,\beta)}{\partial x} = 0$. For instance in the (univariate) quadratic regression model $\eta(x,\beta) = \beta_1 + \beta_2 x + \beta_3 x^2$, we would have to choose $L^T = l^T = (0, -\frac{1}{2\beta_3}, 0)$.

Certain design problems for locating extremal values, however, can be simplified by a suitable reparameterization of the response function. One can then find an exact solution of the problem, conditional on the true value of the parameters, which gives more insight into the properties of the resulting design. Such a solution has been suggested first by Buonaccorsi and Iyer (1986), but the idea was only fully exploited by Fedorov and Müller (1997).

It is obvious that we can rewrite the quadratic regression (3.5) as

$$\eta(x,\tau) = \tau_0 + \tau_{11}(\tau_1 - x_{[1]})^2 + \tau_{12}(\tau_1 - x_{[1]})(\tau_2 - x_{[2]}) + \tau_{22}(\tau_2 - x_{[2]})^2, \quad (3.12)$$

where the parameters τ_1 and τ_2 now denote the location of the extremal point of the polynomial, and $x \in \mathcal{X}$. A change of scale $u_1 = x_{[1]} - \tau_1$ and $u_2 = x_{[2]} - \tau_2$ then leads to a design optimization problem for a second order polynomial on the transformed design space $\mathcal{U} = \{\binom{u_1(x)}{u_2(x)}, x \in \mathcal{X}\}$, which is completely equivalent to the original problem (see Appendix A.3.1).

This new viewpoint has the advantage that the resulting design depends upon the prior guess of the extremum only via the boundaries of the transformed design space, which makes it easy to apply geometric intuition in this setting. As an example consider that we believe that the extremal point is at $(\hat{\tau}_1, \hat{\tau}_2)^T = (0.75, 0.75)$ in $\mathcal{X} = [-1, 1]^2$.

We can now solve the design problem on $\mathcal{U} = [-1.75, 0.25]^2$ and then remap it to the original design space. The resulting D_L-optimum design (with $L^T = \begin{pmatrix} 1 & 0 & 0 & 0 & 0 & 0 \\ 0 & 1 & 0 & 0 & 0 & 0 \end{pmatrix}$) is a 7-point design with measure $\xi^*_{D_L}(0,0) = 0.25$, $\xi^*_{D_L}(1,0) = \xi^*_{D_L}(0,1) = \xi^*_{D_L}(1,1) = 0.208$, and $\xi^*_{D_L}(-1,0) = \xi^*_{D_L}(0,-1) = \xi^*_{D_L}(-1,-1) = 0.046$. Its sensitivity function is given in Figure 3.5 and it is remarkable that it takes its maximum value ($\Phi'[M(\xi^*)] = 2$) also at points $(-1,1)$ and $(1,-1)$, where $\xi^*_{D_L}$ is not supported. If we assume the extremal value closer to the center of $\mathcal{X}$, the D_L-optimum design collapses to a 4-point design, e.g. for $(\hat{\tau}_1, \hat{\tau}_2)^T = (0.5, 0.5)$ we have $\xi^*_{D_L} = \{(0,0), (0,1), (1,0), (1,1)\}$ with equal weights. For the one-dimensional case Fedorov and Müller (1997) study the robustness of the design with respect to the prior guesses of the parameters. Their result indicates that it is less inefficient to wrongly assume the point of extremum closer to the boundary of the design region $\mathcal{X}$.

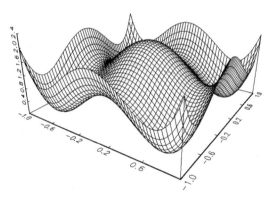

FIGURE 3.5. Sensitivity function (vertical) for the extremal value problem on the unit square.

Designs for Random Coefficient Models (Bayesian Regression)

A common generalization of the basic model (2.1) is to allow the parameters β to be random quantities as well. Specifically, we assume

$$y_j(x) = \eta(x, \beta_j) + \varepsilon_j(x), \quad j = 1, \ldots, J, \qquad (3.13)$$

with
$$E[\beta_j] = \beta, \qquad E[(\beta_j - \beta)(\beta_{j'} - \beta)] = D,$$
and the errors are independent with zero mean and variance σ^2 and independent of β_j. This setup allows for additional model flexibility, which is especially useful for spatio-temporal data as can be seen from the example in the following section.

Fortunately, the situation is covered by the general equivalence theorem (see Gladitz and Pilz (1982) and Fedorov and Müller (1989a)) and in case of D-optimality we have the sensitivity functions

$$\phi_\beta(x, \xi_N) = \dot{\eta}^T(x) \left[\frac{\sigma^2}{N} D^{-1} + M(\xi_N)\right]^{-1} \dot{\eta}(x), \qquad (3.14)$$

and

$$\phi_{\beta_j}(x, \xi_N) = \dot{\eta}^T(x) M^{-1}(\xi_N) \left[\frac{N}{\sigma^2} D + M^{-1}(\xi_N)\right]^{-1} M^{-1}(\xi_N) \dot{\eta}(x), \qquad (3.15)$$

for estimation of β and β_j respectively.

Both functions depend upon the number of observations N and can therefore strictly be meaningful only for exact designs. However, we can interpret the ratio $\frac{\sigma^2}{N}$ as a continuous weighting parameter that combines prior information about the parameters with information from the experiment.

It turns out, that (3.14) coincides with the sensitivity function for a Bayesian estimator in a Gaussian conjugate prior setting. Therefore versions of this function (most frequently those based upon (3.4)) are used to calculate optimum Bayesian designs. In principle any of the optimality criteria (not only D-optimality) given in section 3.2 can be used for the determination of those designs. However, Chaloner (1984) propagates a criterion (**Bayes-A-Optimality**), that might meet the Bayesian philosophy (incorporation of prior information whenever possible) better. Let the aim of the experiment be the estimation of a class of functions $l^T \beta$, with a probability measure π defined on l. Now the expected preposterior risk has to be minimized, i.e.

$$\Phi[M(\xi)] = -\mathrm{tr} E_\pi[ll^T] \left[M(\xi) + \frac{\sigma^2}{N} D^{-1}\right]^{-1}.$$

Bayesian design problems are treated extensively in the seminal monograph by Pilz (1991); for two briefer recent reviews see Chaloner and Verdinelli (1995) and DasGupta (1996).

Another remarkable feature of random coefficient models is that it can be shown (see Appendix A.3.2) that a backward step based upon (3.14), approximately coincides with a heuristic design reduction procedure (cf. Der Megréditchian (1985)). Namely from a set of design points $x_1, \ldots, x_N$ with an observed covariance matrix $\hat{K}$ delete the point x_{i^-}, where

$$i^- = \arg\max_i \hat{K}_{ii}^{-1}. \qquad (3.16)$$

Relationship of this procedure to other more formal methods are revealed in Chapters 4 and 6.

Moving Sensors

In most parts of this monograph it is assumed that the data collection sites (sensors) are stationary, i.e. they are not allowed to change their positions over various points in time. However, in case the observations are assumed to be generated from a distributed parameter system (3.6) it is quite natural to address the question of moving sensors, so that $x_i(t) \in \mathcal{X}$, $t \in [0, T]$. Under mild assumptions the information matrix (3.2) readily generalizes to this case and Rafajłowicz (1986) thus practically reduces the problem to finding a separate optimal design for every time point. This solution has the disadvantage that the induced relocations of the sensors do not necessarily follow continuous paths. This is overcome by Uciński (1999), who parameterizes those paths and applies optimal control techniques.

3.5 Example

The use of *o.d.e.* theory to construct (or improve) air pollution monitoring networks was firstly suggested by Gribik *et al.* (1976) and we will essentially follow their application ideas in this section.

Upper-Austria has a dominant industrial center, the capital Linz, which produces the major part of the SO_2 emissions in the region. We can therefore assume that part of the observed SO_2 concentration at any point in the county is due to a transport phenomenon. One can assume, that the detrended observations $\tilde{y}_t(x_i)$, where local influences have been removed, follow a spatial spreading pattern that is centered at Linz. The form of such a pattern is sometimes (cf. e.g.

Pitovranov et al. (1993)) assumed proportional to a bivariate Gaussian p.d.f., i.e. $E[\tilde{y}_t(x)] =$

$$\eta(x,\beta) = \beta_7 + \beta_6 e^{\left\{\frac{-1}{2(1-\beta_5^2)}\left[\left(\frac{x_{[1]}-\beta_1}{\beta_3}\right)^2 - 2\beta_5\frac{(x_{[1]}-\beta_1)(x_{[2]}-\beta_2)}{\beta_3\beta_4} + \left(\frac{x_{[2]}-\beta_2}{\beta_4}\right)^2\right]\right\}},$$

where β_1 and β_2 denote the location of the peak, β_3, β_4 and β_5 the standard deviations and the correlation coefficient, and β_6 and β_7 are scaling parameters.

A Locally Optimum Design

In order to find a design that is efficient for estimating the parameters β of this model by applying methods that were presented above, we require prior guesses β_0. They can be readily obtained from the existing data. A reasonable choice is

$$\beta_0 = \hat{\beta}_{NLS} = \arg\min_{\beta} \sum_{t=1}^{T}\sum_{i=1}^{N}[y_t(x_i) - \eta(x_i,\beta)]^2, \tag{3.17}$$

which yields $\hat{\beta}_0^T = (0.604, 0.382, 1.873, 0.854, 0.508, 0.547, -0.400)$ for the given data set (Appendix A.1). The peak is thus identified at a location about 15 kilometers east of Linz, which could be due to wind influences.

By substituting $\beta_0 = \hat{\beta}_{NLS}$ it is possible to calculate a D-optimum design for estimating β which is depicted in Figure 3.6. The circles indicate the location of support points and their area is proportional to the respective design measure $\xi(x_i)$. The optimum design spreads itself along the boundaries of the design region, with one central point. Two pairs of points are quite close to each other and could be joined without much loss in efficiency. The design reminds very much of a standard design for second degree polynomials, which is not surprising considering the similar shape of the two models over a wide region of the design space.

A Design for Localizing the Peak

It is simple to devise a design for efficient detection of the peak in model (3.17). Since β_1 and β_2 completely determine the location of the extremum a D_L-optimum design with $L^T = \begin{pmatrix} 1 & 0 & 0 & 0 & 0 & 0 & 0 \\ 0 & 1 & 0 & 0 & 0 & 0 & 0 \end{pmatrix}$

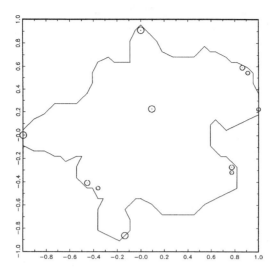

FIGURE 3.6. A locally D-optimum design.

will be the objective. Such a design, based on the same prior guess β_0, spreads its measure in the north-eastern part of the region (around the assumed location of the peak). A graphical representation can be found in Figure 3.7. Here the four central points could again be merged.

A Design for a Random Coefficient Model

Representation (3.13) provides us with a more flexible (and realistic) possibility than (2.26) for modeling the spatio-temporal behavior of the studied phenomenon. We can now treat every day as a particular realization of a random coefficient model stemming from a parameter β_t. To utilize (3.14) and (3.15) we need to replace the unknown σ^2 and D by empirical estimates. Also $M(\xi_N)$ needs to be evaluated at a proper prior guess. For $\hat{\beta}$ (3.17) turns out to be appropriate (cf. Fedorov et al. (1993)) and $\hat{\beta}_t$ is evaluated analogously from the appropriate subset. Instead of σ^2 we use

$$\hat{\sigma}^2 = \frac{1}{T(N-m)} \sum_{t=1}^{T} \sum_{i=1}^{N} [y_t(x_i) - \eta(x_i, \hat{\beta}_t)]^2,$$

and D is replaced by

$$\hat{D} = \frac{1}{T-1} \sum_{t=1}^{T} (\hat{\beta}_t - \hat{\beta})(\hat{\beta}_t - \hat{\beta})^T - \hat{\sigma}^2 M^{-1}.$$

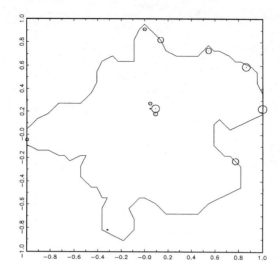

FIGURE 3.7. A D_L-optimum design for localizing the peak.

After these replacements an algorithm based on the sensitivity function (3.15) can be used to construct a D-optimum design for the estimation of β (as well as β_t). The resulting design collapses its measure on basically four points from which two are in regions where no observations have been taken previously. This indicates locations where it is most useful to complement the prior information with additional data.

3.6 Exercises

1. Fit an appropriate nonlinear global model to the data and use the resulting parameter estimates to find a locally optimum design. Vary the estimates a bit and compare the respective resulting designs.

2. Compare the above design to one suitable for estimating a second order polynomial. Are there unexpected differences — if yes, to what could they be due to?

3. Use the reparameterization technique based on model (3.12) to construct a design for estimating the location of maximum concentration in the field. What criterion would one use to find an optimum design for estimating the maximum concentration value?

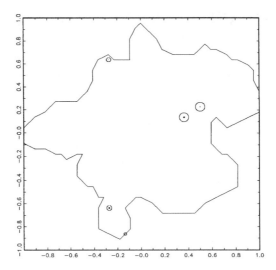

FIGURE 3.8. A D-optimum design for estimating the mean parameter in a random coefficient model - a Bayesian D-optimum design

4. Use a random coefficient formulation of the quadratic model and estimate the corresponding covariance matrix of the parameters. With the resulting estimates construct an optimum design for the mean and the individual parameters with respect to a particular day. Interpret the difference, especially concerning the number of support points.

5. Use the design technique by Der Megréditchian (1985) to reduce the number of sites accordingly. Which design does the result resemble?

3. Fundamentals of Experimental Design

Connections to Other Chapters

In this chapter fundamental methods of optimum design theory were presented as they apply to the uncorrelated error setting. Modifications of these methods for the correlated case including an extension of the definitions of design measures and information matrices can be found in Chapter 5. Chapter 4 gives design methods under less stringent model assumptions. In Chapter 6 principles of *o.d.e.* are applied to construct optimum designs for variogram estimation. The basic algorithmic idea of one-point corrections introduced in this chapter will be used throughout this monograph.

References

Atkinson, A.C., Chaloner, K., Herzberg, A.M. and Juritz, J., (1993). Optimum experimental designs for properties of a compartmental model. *Biometrics*, 49:325–337.

Atkinson, A.C. and Donev, A.N. (1992). *Optimum Experimental Designs*. Oxford Statistical Science Series No.8, Oxford University Press.

Atkinson, A.C. and Haines, L.M. (1996). Designs for nonlinear and generalized linear models. In Gosh, S. and Rao, C.R., editors, *Handbook of Statistics*, volume 13. Elsevier, Amsterdam, 437–475.

Atwood, C.L. (1973). Sequences converging to D-optimal designs of experiments. *The Annals of Statistics*, 1(2):342–352.

Bandemer, H., Bellmann, A., Jung, W., Le Anh Son, Nagel, S., Nagel, W., Näther, W., Pilz, J. and Richter, K. (1977). *Theorie und Anwendung der optimalen Versuchsplanung I*. Akademie Verlag, Berlin.

Box, G.E.P., Hunter, W.G. and Hunter, J.S. (1978). *Statistics for Experimenters. An Introduction to Design, Data Analysis, and Model-Building*. John Wiley & Sons, New York.

Box, G.E.P. and Draper, N.R. (1975). Robust designs. *Biometrika*, 62:347–352.

Box, G.E.P. and Draper, N.R. (1987). *Empirical Model-Building and Response Surfaces*. John Wiley & Sons, New York.

Box, G.E.P. and Lucas, H.L. (1959). Design of experiments in nonlinear situations. *Biometrika*, 46:77–90.

Box, G.E.P. and Wilson, K.B. (1951). On the experimental attainment of optimum conditions (with discussion). *Journal of the Royal Statistical Society, Series B*, 13:1–45.

Buonaccorsi, J.P. and Iyer, H.K. (1986). Optimal designs for ratios of linear combinations in the general linear model. *Journal of Statistical Planning and Inference*, 13:345–356.

Chaloner, K. and Verdinelli, I. (1995). Bayesian experimental design: A review. *Statistical Science*, 10(3):273–304.

Chaloner, K. (1984). Optimal Bayesian experimental design for linear models. *The Annals of Statistics*, 12(1):283–300.

Chaloner, K. (1993). A note on optimal Bayesian design for nonlinear problems. *Journal of Statistical Planning and Inference*, 37:229–235.

Cheng, C. and Li, K. (1987). Optimality criteria in survey sampling. *Biometrika*, 74:337–345.

Chernoff, H. (1953). Locally optimal designs for estimating parameters. *Annals of Mathematical Statistics*, 24:586–602.

Chernoff, H. (1972). *Sequential Analysis and Optimal Design*. Society for Industrial and Applied Mathematics, Philadelphia.

Cook, R.D. and Fedorov, V. (1995). Constrained optimization of experimental design (with discussion). *Statistics*, 26:129–178.

Cook, R.D. and Nachtsheim, C.J. (1980). A comparison of algorithms for constructing exact D-optimal designs. *Technometrics*, 22:315–324.

Cox, D.R. (1958). *Planning of Experiments*. John Wiley and Sons, New York.

Curtis, A. (1999). Optimal experiment design: cross-borehole tomographic examples. *Geophysical Journal International*, 136:637–650.

DasGupta, A. (1996). Review of optimal Bayes designs. In Rao, C.R. and Gosh, S., editors, *Handbook of Statistics*, volume 13. North-Holland.

Der Megréditchian, G. (1985). *Methodes statistiques d'analyse et d'interpolation des champs meteorologiques* (in French). Organisation Meteorologique Mondiale, Geneve.

Dette, H., Heiligers, B. and Studden, W.J. (1995). Minimax designs in linear regression models. *The Annals of Statistics*, 23(1):30–40.

Dette, H. and Haines, L.M. (1994). E-optimal designs for linear and nonlinear models with two parameters. *Biometrika*, 81:739–754.

Elfving, G. (1952). Optimum allocation in linear regression theory. *The Annals of Mathematical Statistics*, 23:255–262.

Fedorov, V.V., Hackl, P. and Müller, W.G. (1993). Estimation and experimental design for second kind regression models. *Informatik, Biometrie und Epidemiologie in Medizin und Biologie*, 3(24):134–151.

Fedorov, V.V., Hackl, P. and Müller, W.G. (1995). Optimal and practicable designs for measuring plaque pH-profiles. *Statistics in Medicine*, 14:2609–2617.

Fedorov, V.V. and Hackl, P. (1997). *Model-Oriented Design of Experiments*, volume 125 of *Lecture Notes in Statistics*. Springer Verlag, New York.

Fedorov, V.V. and Müller, W.G. (1989). Comparison of two approaches in the optimal design of an observation network. *Statistics*, 20(3):339–351.

Fedorov, V.V. and Müller, W.G. (1989). Design of an air-pollution monitoring network. An application of experimental design theory. *Österreichische Zeitschrift für Statistik und Informatik*, 19(1):5–18.

Fedorov, V.V. and Müller, W.G. (1997). A reparametrization view of optimal design for the extremal point in polynomial regression. *Metrika*, 46:147–157.

Fedorov, V.V. (1972). *Theory of Optimal Experiments*. Academic Press, New York.

Fedorov, V.V. (1980). Convex design theory. *Mathematische Operationsforschung und Statistik, Series Statistics*, 11(3):403–413.

Firth, D. and Hinde, J. (1997). Parameter neutral optimum design for nonlinear models. *Journal of the Royal Statistical Society, Series B*, 59(4):799–811.

Fisher, R.A. (1935). *The Design of Experiments*. Oliver & Boyd Ltd., Edinburgh.

Ford, I., Titterington, D.M. and Kitsos, C.P. (1989). Recent advances in nonlinear experimental design. *Technometrics*, 31(1):49–60.

Ford, I., Torsney, B. and Wu, C.F.J. (1992). The use of a canonical form in the construction of locally optimal designs for non-linear problems. *Journal of the Royal Statistical Society, Series B*, 54(2):569–583.

Ford, I. and Silvey, S.D. (1980). A sequentially constructed design for estimating a nonlinear parametric function. *Biometrika*, 67:381–388.

Gaffke, N. and Heiligers, B. (1995). Algorithms for optimal design with application to multiple polynomial regression. *Metrika*, 42:173–190.

Gaffke, N. and Mathar, R. (1992). On a class of algorithms from experimental design theory. *Optimization*, 24:91–126.

Gladitz, J. and Pilz, J. (1982). Construction of optimal designs in random coefficient models. *Mathematische Operationsforschung und Statistik, Series Statistics*, 13(3):371–385.

Gribik, P., Kortanek, K. and Sweigart, J. (1976). Designing a regional air pollution monitoring network: an appraisal of a regression experimental design approach. In *Proceedings of the Conference on Environmental Modelling and Simulation*. EPA, 86–91.

Haines, L.M. (1987). The application of the annealing algorithm to the construction of exact optimal designs for linear regression models. *Technometrics*, 29:439–447.

Haines, L.M. (1993). Optimal design for nonlinear regression models. *Communications in Statistics, Theory & Methods*, 22:1613–1627.

Hamilton, D.C. and Watts, D.G. (1985). A quadratic design criterion for precise estimation in nonlinear regression models. *Technometrics*, 27(3):241–250.

Hedayat, A. (1981). Study of optimality criteria in design of experiments. *Statistics and Related Topics*, 39–56.

Kiefer, J. and Wolfowitz, J. (1960). The equivalence of two extremum problems. *Canadian Journal of Mathematics*, 14:363–366.

Kiefer, J. (1959). Optimal experimental designs (with discussion). *Journal of the Royal Statistical Society, Series B*, 272–319.

Kiefer, J. (1974). General equivalence theory for optimum designs (approximate theory). *Annals of Statistics*, 2:849–879.

Kiefer, J. (1975). Optimal design: Variation in structure and performance under change of criterion. *Biometrika*, 62:277–288.

Kiefer, J. (1985). *Collected Papers*. Springer, New York.

Martin, R.J. (1996). Spatial experimental design. In Rao, C.R. and Gosh, S., editors, *Handbook of Statistics*, volume 13. North-Holland, 477–513.

Mirnazari, M.T. and Welch, W.J. (1994). Criterion-robust optimal design. TR STAT-94-03, Department of Statistics and Actuarial Science, University of Waterloo.

Mitchell, T.J. (1974). An algorithm for the construction of "D-optimal" experimental designs. *Technometrics*, 16(2):203–210.

Molchanov, I. and Zuyev, S. (2000). Variational calculus in space of measures and optimal design. In Atkinson, A.C., Bogacka, B., and Zhigljavsky, A.A., editors, *Optimum Design 2000*. Kluwer, 79–90.

Müller, W.G. and Pötscher, B.M. (1992). Batch sequential design for a nonlinear estimation problem. In Fedorov, V.V., Müller, W.G. and Vuchkov, I., editors, *Model-Oriented Data Analysis 2*, Physica Verlag, Heidelberg.

Müller, W.G. (1995). Contribution to discussion of 'Constrained Optimization of Experimental Design' by D. Cook and V. Fedorov. *Statistics*, 26:166–167.

Myers, R.H. and Montgomery, D.C. (1995). *Response surface methodology: Process and product optimization using experiments.* J.Wiley, New York.

Nguyen, N.K. and Miller, A.J. (1992). A review of some exchange algorithms for constructing discrete D-optimal designs. *Computational Statistics and Data Analysis*, 14:489–498.

Pázman, A. and Pronzato, L. (1992). Nonlinear experimental design based on the distribution of estimators. *Journal of Statistical Planning and Inference*, 33:385–402.

Pázman, A. (1986). *Foundations of Optimum Experimental Design*. Mathematics and Its Applications. D.Reidel, Dordrecht.

Pilz, J. (1991). *Bayesian Estimation and Experimental Design in Linear Regression Models*. Wiley Series in Probability and Mathematical Statistics, New York.

Pirsig, R.M. (1974). *Zen and the Art of Motorcycle Maintenance*. William Morrow and Company, Inc., New York.

Pitovranov, S.E., Fedorov, V.V. and Edwards, L.L. (1993). Optimal sampler siting for atmospheric tracer experiments taking into account uncertainties in the wind field. *Atmospheric Enviroment*, 27A(7):1053–1059.

Pukelsheim, F. and Rieder, S. (1992). Efficient rounding of approximate designs. *Biometrika*, 79:763–770.

Pukelsheim, F. (1993). *Optimal Design of Experiments*. John Wiley & Sons, Inc., New York.

Quereshi, Z.H., Ng, T.S. and Goodwin, G.C. (1980). Optimum experimental design for identification of distributed parameter systems. *International Journal of Control*, 31(1):21–29.

Rafajłowicz, E. (1986). Optimum choice of moving sensor trajectories for distributed parameter system identification. *International Journal of Control*, 43(5):1441–1451.

Shah, K.R. and Sinha, B.K. (1989). On the choice of optimality criteria comparing statistical designs. *Canadian Journal of Statistics*, 17:345–348.

Silvey, S.D., Titterington, D.M. and Torsney, B. (1978). An algorithm for optimal designs on a finite design space. *Communications in Statistics, A* 7:1379–1389.

Silvey, S.D. (1980). *Optimal Design.* Chapman and Hall, London.

Torsney, B. (1977). Contribution to discussion of 'Maximum likelihood from incomplete data via the EM-algorithm' by Dempster, A.P., Laird, N.M. and Rubin, D.B., *Journal of the Royal Statistical Society, Series B*, 39:26–27.

Uciński, D. (1999). *Measurement Optimization for Parameter Estimation in Distributed Systems.* Technical University Press, Zielona Gora.

White, L.V. (1973). An extension to the general equivalence theorem for nonlinear models. *Biometrika*, 60:345–348.

Whittle, P. (1973). Some general points in the theory of optimal design in a nonlinear situation. *Journal of the Royal Statistical Society, Series B*, 35:123–130.

Wiens, D.P. (1992). Minimax designs for approximately linear regression. *Journal of Statistical Planning and Inference*, 31:353–371.

Wilhelm, A. (1997). Computing optimal designs by bundle trust methods. *Statistica Sinica*, 7(3):739–753.

Wynn, H.P. (1970). The sequential generation of D-optimum experimental designs. *Annals of Mathematical Statistics*, 41:1655–1664.

4
Exploratory Designs

> *"In Wien geht man ins Café, um sich zurückzuziehen, und jeder setzt sich, inselbildend, soweit wie möglich von jedem anderen — [...]"*
> von Doderer (1951)

Optimum designs as constructed by the methods given in Chapter 3 (by their critics sometimes belittlingly called alphabetic optimality designs) have been heavily critized in the literature (see e.g. Box and Draper (1987)), mainly for concentrating too much on a single objective, the design criterion, and for not taking into account other requirements of well-designed experiments. Box and Draper (1975) provide a list of 14 such requirements. Some of these can be directly incorporated in the design criterion ('assume that the fitted value $\hat{y}(x)$ be as close as possible to the true value $\eta(x)$'), others do not make sense in the spatial setting ('not require an impractically large number of levels of the predictor values').

Some of the points made in these criticisms, however, revolve around the fact that the designs derived from o.d.e. principles critically rely on the assumption of perfect knowledge of the (form of the) response and that they may be very sensitive to deviations from this basic assumption. An optimum design for an m-th degree polynomial, for instance, typically consists of only m support points and thus identification of a higher order model becomes impossible. Since in spatial statistics the assumed models are often only rough approximations of complex phenomena the problem of robustness against model misspecification is of great relevance.

In principle there are two remedies for that deficiency in the approaches discussed so far. First, to directly implement aspects that make the approach more robust into the design criterion. This avenue is taken by Box and Draper (1959), who suggest instead of maximizing a function $\Phi[\cdot]$ based upon the information matrix to rather minimize the assumed integrated mean squared error (IMSE) under misspecification, i.e.

$$\mathcal{J}(\xi_N) = -\int_{\mathcal{X}} E[y(x) - \hat{y}(x|\xi_N)]^2 dx,$$

where $\hat{y}(\cdot|\xi_N)$ denotes the predicted response based upon the observations from a design ξ_N. The corresponding design principle is sometimes called **J-Optimality**. Furthermore, they propagate to drop the variance term completely and to solely use bias minimization as the main criterion. An investigation of this criterion in the two-dimensional context is given in Evans and Manson (1978). The logic behind this approach is not flawless. For the evaluation of the IMSE we need to assume a 'true' model that we have violated by the choice of our response function. If such a true model is indeed known, why not to use it directly in the spirit of Chapter 3? Another objection against bias minimizing designs is of course, that 'no design which takes a finite number of observations can protect against all possible forms of bias' (Chang and Notz (1996)). In Chapter 6 some alternative, more compromising methods that may be used for direct inclusion of various objectives in the design criterion are presented.

The second general approach to robust design is to relax the rigidity of prior assumptions (such as the knowledge of the 'true' response function). Designs resulting from such approaches can be called exploratory designs, in the sense that they provide enough freedom for allowing to explore the structure of the processes under study in the classical model-building spirit. Such designs could be adopted in an initial data collection step, as a starting design in design improvement algorithms, or they could be combined with optimum designs for better overall performance (see again Chapter 6).

There are several good reasons for initially using an exploratory design with n_0 sites in $\mathcal{X}$. First, data from such a design can be used to carry out the initial stage of a geostatistical analysis, that is, to determine an appropriate trend $\eta(x,\beta)$ in (2.1) and to check that the error process $\varepsilon(\cdot)$ satisfies the intrinsic stationarity assumption. Second, the data provide initial estimates of β and of θ; the former is useful for the nonparametric estimation of the variogram $\gamma(h)$ when $\eta(x,\cdot)$ is not constant, the latter is useful because the optimal choice of the remaining $n - n_0$ sites generally depends on θ. Third, if the design is chosen wisely, the initial data can be used to check for anisotropy. Fourth, a properly chosen starting design (e.g. $n_0 > m$) ensures the regularity of the (information) matrices involved in further computations. The first three sections of this chapter are concerned with how such an (initial) exploratory design may be constructed reasonably.

Another aspect for design construction that is interconnected with the robustness issue is, that if we are not sure about the appropriateness of our model, we would like the design to 'give good detectability of lack of fit' (Box and Draper (1975)). This includes to be able to differentiate between models from a list of candidates, a problem that is treated by so-called model discrimination designs presented in Section 4.4.

4.1 Deterministic and Random Sampling

Of course, in some situations the statistician plays no role in the selection of the initial design; some data have already been collected, perhaps where it was most convenient to do so, and the design problem is that of determining how to augment best the existing design. In other situations (e.g. soil surveys), however, one can build the design 'from scratch'.

In the latter case, an exploratory design that satisfactorily accomplishes the goals described above is a regular grid, such as a square or triangular grid. Such a design is well suited for determining an appropriate response function, for checking whether assumptions about the errors are reasonably well satisfied, and it yields lags in numerous directions for checking for anisotropy. There seems little by which to choose between a square grid or a triangular grid for purposes of model discrimination or/and parameter estimation. It is worth noting, however, that the square grid is slightly more convenient from a practical standpoint (easier site determination) and that the triangular grid seems marginally more efficient for purposes of prediction (cf. Matern (1960), Yfantis et al. (1987)).

Bellhouse and Herzberg (1984) have compared D-optimum designs and uniform grids (in a one-dimensional setup) and they come to the conclusion that (depending upon the model) the prediction of $y(x)$ can in certain regions actually be improved by regular grids. In the case of second order polynomials on $[-1, 1]$ for instance they identify this region as $|x| < 0.58$. A comparison in a multi-dimensional setup (including correlations) can be found in Herzberg and Huda (1981).

For higher dimensional problems Bates et al. (1996) recommend the use of (nonrectangular) lattices rather than grids (they also reveal connections to D-optimality). In the two-dimensional setup (on the unit square $[-1, 1]^2$) the Fibonacci lattice

$$x_{i1} = 2\left\{\frac{i + 0.5}{n_0} \mod 1\right\} - 1, \qquad (4.1)$$

$$x_{i2} = 2\left\{\frac{\zeta_{k-1}(i-1) + 0.5}{n_0} \mod 1\right\} - 1,$$

$$\zeta_1 = \zeta_2 = 1, \quad \zeta_k = \zeta_{k-1} + \zeta_{k-2}, \quad \text{for } k \geq 3$$

for $i = 1, \ldots, n_0 = \zeta_k$ (see Koehler and Owen (1996)) proved to be useful. The advantage of lattices is that their projection on lower dimensions covers the design region more or less uniformly; for an example see Figure 4.1. Adaptations to irregular design regions may not be straightforward, but good enough approximations will suffice. This is not the case for many other systematic designs that are frequently proposed in the literature, such as central composite designs (Box and Wilson (1951)), the construction of which relies on the symmetry of the design region, which is rarely the case in our context.

It is evident that randomization can be helpful for making designs more robust. On a finite grid $\bar{\mathcal{X}}$ with $\mathcal{N}$ candidate points we can think of randomization as drawing a single design ξ_{n_0} from Ξ according to a prespecified probability distribution $\pi(\xi)$. The uniform distribution $\pi(\cdot) = \binom{\mathcal{N}}{n_0}^{-1}$ then corresponds to simple random sampling and more refined schemes (e.g. stratified random sampling, see Fedorov and Hackl (1994)) can be devised by altering $\pi(\cdot)$.

A comparison between deterministic selection and random sampling is hard to make, since for a finite sample it is evident that for any single purpose it is possible to find a deterministic design that outperforms

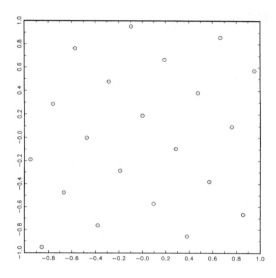

FIGURE 4.1. A 21 point Fibonacci lattice on the unit square.

random sampling. A comparison attempt by simulation techniques under poor prior knowledge is made by McArthur (1987). He comes to the not very surprising conclusion that for the estimation of the average level of a locally concentrated pollutant (stratified) regular sampling is far superior to any random sampling scheme.

More information about the relative usefulness of the discussed methods can be gained from an analysis of their asymptotic performance. Such an analysis for the one-dimensional case is reviewed in detail by Cambanis (1985), who also gives references for extended settings. If the comparison is based upon the convergence rate of the mean squared error of the parameters β it turns out that in the two-dimensional case random and deterministic sampling strategies perform equally well, both having a rate n^{-1}. The asymptotic constants of the m.s.e. of asymptotically optimum designs (see also section 5.3) are generally larger than that of regular sampling designs. A slight improvement is offered by stratification, depending upon the smoothness of the covariance function $c(\cdot,\cdot)$. There is an indication however, see Ylvisaker (1975), that deterministic product designs can give an even better rate of convergence.

Deterministic designs that were developed specifically for the purpose of variogram estimation which require very few prior assumptions and are therefore suitable as exploratory designs for problems treated in Chapter 6 were recently given by Pettitt and McBratney (1993).

4.2 Space Filling Designs

Designs selected with only parameter estimation in mind have also been criticized to leave large unsampled 'holes' in the design region. Particularly in the absence of prior knowledge about the structure of the response we would want that a design achieves good overall coverage of $\mathcal{X}$. Many of the methods presented in the previous section seem to ensure a reasonable degree of overall coverage of the study area. However, there have been claims (see e.g. Fang and Wang (1994)), that the efficiency (with respect to coverage) of these methods may be poor, when the number of design points is small.

To allow for comparisons between designs in the above respect Fang (1980) introduced some formal criteria, the so-called discrepancy

$$\mathcal{D}(\xi_n) = \max_{x \in \mathcal{X}} |F_n(x) - U(x)|,$$

amongst them. Here $U(x)$ is the c.d.f. of a uniform distribution on $\mathcal{X}$ and $F_n(x)$ denotes the empirical c.d.f. for the support S_{ξ_n}. The discrepancy by this definition is just the Kolmogorov-Smirnov test statistic for the goodness-of-fit test for a uniform distribution. Based upon this definition Fang and Wang (1994) call a design that guarantees

$$\mathcal{D}(\xi_n) = o(n^{\frac{1}{2}}) \qquad \text{for } n \to \infty$$

uniformly scattered and thus suggest to find 'optimum' designs according to

$$\xi_n^* = \arg\min_{\xi_n} \mathcal{D}(\xi_n),$$

which they term the **U-criterion**. It turns out that for certain choices of n lattice designs given in (4.1) are U-optimum. Those lattice designs are also D-optimum for some specific Fourier regressions and these and other connections are explored by Riccomagno et al. (1997).

A different approach to space-filling is taken by Johnson et al. (1990). They propose minimax and maximin distance designs, where the latter are defined as

$$\xi_n^* = \arg\max_{\xi_n} \min_{x_i, x_{i'} \in S_{\xi_n}} h_{ii'}. \tag{4.2}$$

Such designs have been originally developed for the basic random field model (2.1) and it was shown that, if a covariance function

$$c(h, \cdot) = [c'(h, \cdot)]^\kappa$$

is used with $c'(h, \cdot)$ being itself a decreasing covariance function, a design according to (4.2) coincides with

$$\xi_n^* = \arg\max_{\xi_n} |C|,$$

if $\kappa \to \infty$. This is an optimum design rule, which is intuitively in accordance with the heuristic procedure (3.16), which is based upon the empirical counterpart of C; further relations are given in Section 5. A comparison with designs from Chapter 3 for two-level factorial experiments can be found in John et al. (1995).

Maximin distance designs, even for small sample problems, may be difficult to construct (especially when the design region is irregular) and can have multiple structurally different solutions, see e.g. Figure 4.2 for two 6-point maximin distance designs on a circle. Tobias (1995) writes that "space-filling designs can be very time-consuming to produce; search times of days are not uncommon for realistic candidate regions of interest." He and others, e.g. Royle and Nychka (1998), suggest using heuristic point exchange algorithms, whereas Morris and Mitchell (1995) give an extended definition of (4.2) to avoid multiple global optima and employ a simulated annealing algorithm for the computation of their optimum designs.

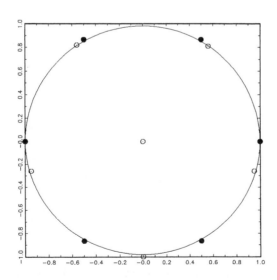

FIGURE 4.2. Two 6-point maximin distance designs on the unit circle (small circles and small solid circles represent sites).

A very different technique is proposed by Nychka et al. (1996). They reformulate the problem into the classical one of selecting regressor subsets in linear regression and employ standard procedures from statistical packages for its solution. Although the resulting designs were created with estimating the average of the random field in mind, their reported coverage properties are satisfactory as well.

'Coffee-house' Designs

To simplify the computations let us instead consider designs that are generated as follows:

a) First the two points

$$S_{\xi_2} = \{x^{(1)}, x^{(2)}\} = \arg\max_{x_i, x_{i'} \in \mathcal{X}} h_{ii'}$$

with maximal distance among all pairs in the design region are found.

b) Then the design is subsequently supplemented such that $S_{\xi_{i+1}} = S_{\xi_i} \cup \{x^{(i+1)}\}$ by

$$x^{(i+1)} = \arg\max_{x_0 \in \mathcal{X}} \min_{x_i \in S_\xi} h_i,$$

where $h_i = \|x_i - x_0\|$, the point which maximizes the minimal distance to all points in S_{ξ_i}, until the desired number of sites n is reached.

We may call designs constructed by this rule 'coffee-house' designs for the similar way in which customers select their tables in a coffee-house.

The above construction algorithm is simple and quick, and the 'coffee-house' designs seem to (at least asymptotically) share some desirable properties of maximin distance designs. Not only

$$\lim_{n\to\infty} P(\max_{x_i^*} \min_{x_{i'}} \|x_i^* - x_{i'}\| < \epsilon) = 1$$

does hold (for an arbitrarily small positive number ϵ and x_i^* being the points in a maximin distance design), which is also true for simple random sampling. Moreover, the ratio $\frac{h_{\min}}{h_{\min}^*}$ of the minimum distance in a 'coffee-house' and the one in a maximin distance design, although not tending to 1, seems to be bounded by below. In the one-dimensional case, for instance, this bound is simply

$$\frac{h_{\min}}{h^*_{\min}} \geq \frac{n-1}{2(n-2)} > \frac{1}{2},$$

i.e. the minimum distance in a coffee-house design is in the worst case twice as large as the optimum (see Appendix A.4.1).

Maximin distance designs, especially when defined on hypercubic regions, usually do not exhibit good overall coverage, when projected onto lower dimensional representations of the design space. For this reason Morris and Mitchell (1995) have suggested to searching for maximin distance designs within a class of Latin hypercube arrangements. Specifically they only allow for designs for which each column of the design matrix contains some permutation of the elements of the set $\{0, \frac{1}{n-1}, \frac{2}{n-1}, \ldots, 1\}$, and they consequently call their resulting designs Minimax distance Latin hypercube (MmLh) designs. An adaptation of 'coffee-house' designs for the same purpose and a comparison to MmLh designs is given by Müller (2000).

4.3 Designs for Local Regression

If, in contrary to previous subsections, little is known about the structure of the process generating model, one has to confine oneself to so-called nonparametric regression approaches, which afford much less assumptions (just about smoothness and differentiability) than the discussed parametric methods. In Section 2.3 one such approach, the moving local regression, was presented as an alternative or complimentary tool for spatial trend estimation.

From the properties of this method it is clear that (depending upon the locality of the weighting scheme) it can be 'nonparameterically' applied in the first stage of a study, in an exploratory analysis to find the general shape of the underlying function. On the other hand if it is impossible to justify stronger assumptions local regression can also be directly used for trend estimation. It is therefore natural to expect that an optimum design corresponding to a local regression problem will give a compromise between optimum designs from Chapter 3 and purely exploratory designs given in Section 4.1.

To make the design problem mathematically more easily tractable reformulate it slightly: At points of interest $\mathcal{Z} = (_1x, \ldots, _jx, \ldots, _qx)$

(possibly at every point of $\bar{\mathcal{X}}$) the unknown response $\eta(\cdot)$ needs to be fitted by local regression (smoothed). The question is again where to take observations (at which x_i's) within $\mathcal{X}$ (which here in principle may also be infinitely expanded) in order to estimate $\eta(_jx)$ in a most efficient way using the scheme (2.12).

We thus need to find a measure of information $\mathcal{J}(\cdot)$ that comprises the performance of the estimated response for all points $(_1x,\ldots,_qx)$. The total (deterministic as well as stochastic) deviation of any estimate should be reflected by such a measure. For expositional simplicity only the minimization of the stochastic deviation (the variance) will be regarded here, under the assumption, that $\hat{\eta}(_jx)$ is an unbiased estimator, which of course holds only in special cases. Nevertheless it seems justified to derive further properties still keeping this assumption, since the amount of the bias can to some extent be governed by the weight function. The amount of control over this issue is investigated in a simulation experiment by Müller (1992). When the true response function (or a more complete approximation) can be assumed, one can adopt a similar approach as given below based upon mean squared error minimization (i.e. including the bias term), see Fedorov et al. (1999). A sequential design optimization algorithm for the same setting is proposed by Cheng et al. (1998).

If one now looks at the presentation $\hat{y} = Ly$ of the estimator, one could guess that it is reasonable to use a scalar function of $\text{Cov}(\hat{\eta}) = L\text{Cov}(y)L^T$ as the criterion of the design problem (with $L^T = (L_1,\ldots,L_q)$). Since the D-criterion in this context would be cumbersome to handle Fedorov (1989) indicates that we can choose a weighted sum of the variances of the estimates $\hat{\eta}(_jx)$ as a sensible objective function $\mathcal{J}(\cdot)$ (a generalized A-criterion).

The optimum design is then given by

$$\xi_N^* = \arg\min_{\xi_N} \sum_{j=1}^{q} w_j \text{Var}(\hat{\eta}(_jx)), \qquad (4.3)$$

where the scalar weights w_j reflect the importance of $_jx$, the variances stemming from the diagonal of $\text{Cov}(\hat{\eta})$.

Applying standard weighted least squares techniques (see Appendix A.4.2) the variance of an estimator can be easily expressed as:

$$\text{Var}(\hat{\eta}(_jx)) = \text{tr}O(X_j^T \text{diag}(w_j)X_j)^{-1} = \text{tr}OM_j^{-1}(\xi_N), \qquad (4.4)$$

where O is a null matrix except $O_{11} = 1$ and $X_j^T = \begin{pmatrix} 1 & \cdots & 1 \\ h_{1j} & \cdots & h_{Nj} \end{pmatrix}$.

Using (4.4) the optimization criterion can directly be formulated as

$$\mathcal{J}(\xi_N) = -\operatorname{tr}\sum_{j=1}^{q} W_j M_j^{-1}(\xi_N), \qquad (4.5)$$

where $W_j = w_j O$. Following the methodology from Chapter 3, we find (see Appendix A.4.3) that maximizing the optimality criterion (4.5) corresponds to a sensitivity function

$$\phi(x_i, \xi) = \sum_{j=1}^{q} w(h_{ij}) X_{ij}^T M_j^{-1}(\xi) W_j M_j^{-1}(\xi) X_{ij} \leq \operatorname{tr}\sum_{j=1}^{q} W_j M_j^{-1}(\xi), \qquad (4.6)$$

where X_{ij} denotes a particular row from X_j.

The above sensitivity function was used by Müller (1996) to generate optimum designs for a great variety of local responses, weighting functions, experimental regions and sets of interest points. It turned out that optimum designs resulting from (4.6) are indeed compromise designs in the discussed sense. If the locality is chosen very narrow, the optimum design will typically spread its points between the interest set $\mathcal{Z}$(similar to space-filling designs). The more the locality is increased, the more the local regression designs tend to the classical designs from Chapter 3.

Designs for Other Smoothers

Note that the use of local regression is more common in environmental statistics than in other application fields (see e.g. Cleveland and McRae (1988)). Nevertheless, there are several other smoothers, that could alternatively be used and their respective optimum designs are expected to have similar properties.

For his class of kernel regression estimates (which is equivalent to a specially restricted form of local regression) Müller (1984) and Müller (1988) for instance also address the problem of optimum design. There the integrated mean squared error (no special points of interest) is used as the criterion function and the optimum design

$$\xi^*(x) = \frac{\sigma(x)}{\int_{\mathcal{X}} \sigma(u) du}$$

where $\sigma(\cdot)$ denotes a smooth variance function, is found. A related method for fitting segmented polynomial regression models that in principle is a non-moving version, or a spline-estimator without differentiability conditions, is fully described by Park (1978). Design problems for spline-estimators are thoroughly treated by Wahba (1992). An entirely Bayesian approach to localized regression with a consequent analysis of optimum design is presented by O'Hagan (1978).

4.4 Model Discriminating Designs

Optimum designs that were constructed exclusively for parameter estimation, i.e. such as discussed in Chapter 3, often do not allow us to discriminate between several candidate models (at least not between those with equal number of parameters), since the design measure is typically concentrated at m design points. With the goal of discriminating between two rival models in mind Atkinson and Fedorov (1975a) have suggested another design criterion, the so-called **T-optimality** criterion. They suggest to choose a design such that

$$\Delta(\xi_N) = \min_{\beta \in \Omega} \sum_{x \in S_{\xi_N}} \{\eta_+(x, \beta_+) - \eta(x, \beta)\}^2 \xi(x), \qquad (4.7)$$

i.e. some kind of lack of fit measure, is maximized, where $\eta_+(\cdot)$ denotes the assumed true model and $\eta(\cdot)$ the rival model. They formulate an equivalence theorem which implies that for approximate (nonsingular) designs the corresponding sensitivity function can be derived as

$$\phi(x, \xi) = \{\eta_+(x) - \eta(x, \beta_\xi)\}^2.$$

Here β_ξ denotes the argument of the minimization in (4.7). In the case of (entirely distinct) linear responses, uncorrelated errors, and $\Omega = \mathbb{R}^m$ it is easy to see (Appendix A.4.4) that

$$\Delta(\xi_N) = N^{-1} \beta_+^T \{X_+^T X_+ - X_+^T X (X^T X)^{-1} X^T X_+\} \beta_+,$$

which is proportional to the noncentrality parameter of the χ^2-distribution in the test for lack of fit of model $\eta(\cdot)$. The design matrices X_+ and X correspond to the assumed true and the rival model, respectively, and β_+ denotes the true values of the parameter in the true model.

From the above presentation it is evident that T-optimality is defined by a local criterion, but local not only with respect to the values of the parameters, but also with respect to which of the rival models is actually the true one. One way out of this dilemma is the sequential algorithm proposed by Atkinson and Fedorov (1975a), where subsequently $\phi(\cdot)$ for both candidates is maximized and a new observation to update the estimates of β_ξ is gathered. This procedure asymptotically leads to T-optimum design under the true model. To allow for nonsequential designs a Bayesian version of this procedure that optimizes a weighted average of the two corresponding T-criteria was suggested by Ponce de Leon and Atkinson (1991). A combination of those two approaches is given by Müller and Ponce de Leon (1996). An entirely Bayesian view of the problem is adopted by Felsenstein (1992).

Note that when the two candidate models are nested, the T-criterion is of little help (since $\Delta(\xi) \equiv 0$), unless we put some restrictions on the parameters β. The same approach is taken by Fedorov and Khabarov (1986) to avoid having to specify an assumed true model. They modify the criterion (4.7) to

$$\min_{\beta:|\eta(x,\beta)|>0, x\in\mathcal{X}} \sum_{x\in S_{\xi_N}} \eta^2(x,\beta)\xi(x),$$

where now $\eta(x,\beta) = \eta_1(x,\beta) - \eta_2(x,\beta)$ denotes the (extended) difference model, i.e. they allow only for models with separate response over the whole region $\mathcal{X}$. Amongst other equalities they find that for approximate designs the above criterion is equivalent to D-optimality in the extended model and thus the application of numerical procedures from Section 3.3 becomes straightforward.

In the case of more than two rival models $\eta_1(\cdot), \ldots, \eta_q(\cdot), \eta_+$ it was the proposal by Atkinson and Fedorov (1975b) to find

$$\xi_N^* = \arg\max_{\xi_N} \min_j \Delta_j(\xi_N),$$

where $\Delta_j(\cdot)$ denotes the respective lack of fit measure of the j-th model. In a sequential procedure this requires an additional step, namely to rank the candidate models by their residual sum of squares and then to proceed by comparing the two lowest ranked. Note that this procedure is not guaranteed to converge to the T-optimum, a detailed account of this feature is given in Atkinson and Donev (1992), Chapter 20.

Pure model discriminating designs have the disadvantage that although they may provide very efficient information about which is the true response they may be quite poor for estimating the parameters of this response. Several methods to incorporate both aspects in the construction of the design were employed in a simple example by Pukelsheim and Rosenberger (1993). For some general ideas about these methods see Chapter 6. The same problem is treated by Borth (1975), who utilizes the total entropy as the design criterion (cf. also Chapter 6).

4.5 Example

A Space-filling Design

Both a 17-point maximin distance and a 'coffee-house' design (Figure 4.3) were calculated for the 45×45 grid representation of Upper-Austria. It turned out that the minimum distances were very close, $h^*_{min} = 0.369$ and $h_{min} = 0.361$ respectively and thus also the design constructed by the much simpler and quicker algorithm ensures reasonable overall coverage of the region.

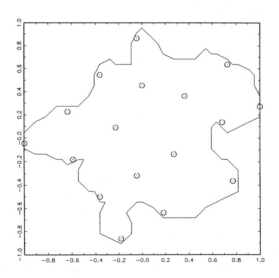

FIGURE 4.3. A 17-point 'coffee-house' design.

A Design for Local Regression

The discussed problem of location of air-pollution monitoring stations can be viewed in the framework of nonparametric regression, when an additional flexibility in the random field model is required. To be specific, if one defines sensible points of interest, i.e. those locations where the pollutant is desired to be estimated most exactly, one can immediately use the local regression technique for estimation of the SO_2 concentration. This will allow the average concentration to form a nonlinear surface over the given region. The methods for optimum design then help to find the location of stations, so that the predictions can be made most precisely.

Let us assume that the set of interest points coincides with $\bar{\mathcal{X}}$, i.e. we desire an equally good estimation at every point in the design region. We now require a prior estimate of $\lambda(\cdot)$ (not of β anymore, since the local model is assumed to be a polynomial). For a local linear response and the tricube weight function in Section 2.5 we have estimated the parameter that identifies $\lambda(\cdot)$ by $\hat{\kappa} = 1.721$, i.e. a weight function with very widespread locality. The optimum design can now be easily calculated with numerical optimization procedures as given in Section 3.3 based upon the sensitivity function (4.6). Not surprisingly the resulting design does not differ much from the corresponding designs from Chapter 3 (for 'global' linearity), i.e. concentrating its measure at just a few points on the boundary of the region. However, if the prior guess $\hat{\kappa}$ is reduced, see e.g. Figure 4.4, the design indeed spreads out more throughout the design region, very similar to a space-filling design.

The points of interest need neither coincide with $\bar{\mathcal{X}}$, nor be equally weighted. A more natural analysis where they were chosen at the Upper-Austrian cities with more than 10000 inhabitants and their influence was weighted with the city population size is given in Müller (1992).

A Design for Model Discrimination

A reasonable alternative model to the bivariate Gaussian p.d.f. for a trend surface with a single emittent source is the polynomial of second order. The extended difference model consists of 11 parameters (the intercepts cancel out) and one can apply D-optimality and a corresponding numerical algorithm to yield the optimum model discrimination design measure in Figure 4.5. Most of the measure is again put

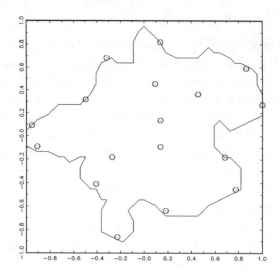

FIGURE 4.4. A design for local regression with a tricube weight function and $\hat{\kappa} = 0.7$.

on the boundary of the region, where the two functions are expected to differ the most. Therefore it is likely that this design performs also well for parameter estimation.

4.6 Exercises

Perform all analyses for this chapter on both the original and the transformed design space and compare the respective results.

1. Construct a 'coffee-house' design with 36 points. Identify points a simple shift of which will improve the design criterion.

2. Construct a couple of random and/or regular designs and compare their coverage properties.

3. For local linear regression find the respective optimum design by using sensitivity function (4.6). How good are its space-filling properties? Vary the set of points of interest and the locality.

4. Assume a true global model (from an earlier chapter) and employ the technique from Fedorov et al. (1999) for finding an optimum local linear regression design. How does the result differ from the above?

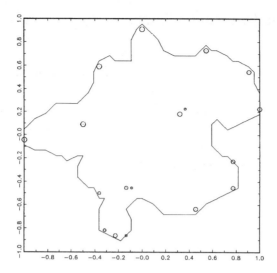

FIGURE 4.5. A design for discriminating a second order polynomial from a Gaussian p.d.f.

5. Contrast the global model with a second order polynomial and calculate the optimum discriminating design. Can this design be also efficiently used for other purposes?

Connections to Other Chapters

In this section designs were presented for the purpose of specifying the form of the response function rather than parameter estimation. The rigid assumptions from Chapter 3 on the model structure were either relaxed, or several candidate models were allowed. Many relationships to parameter estimation designs were indicated. In Chapter 6 a different approach will be discussed, where robustness requirements can enter directly into the design criterion.

References

Atkinson, A.C. and Donev, A.N. (1992). *Optimum Experimental Designs*. Oxford Statistical Science Series No.8, Oxford University Press.

Atkinson, A.C. and Fedorov, V.V. (1975). The design of experiments for discriminating between two rival models. *Biometrika*, 62:57–70.

Atkinson, A.C. and Fedorov, V.V. (1975). Optimal design: Experiments for discriminating between several models. *Biometrika*, 62(2):289–303.

Bates, R.A., Buck, R.J., Riccomagno, E. and Wynn, H.P. (1996). Experimental design and observation for large systems. *Journal of the Royal Statistical Society, Series B*, 58(1):77–94.

Bellhouse, D.R. and Herzberg, A.M. (1984). Equally spaced design points in polynomial regression: a comparison of systematic sampling methods with the optimal design of experiments. *The Canadian Journal of Statistics*, 12(2):77–90.

Borth, D.M. (1975). A total entropy criterion for the dual problem of model discrimination and parameter estimation. *Journal of the Royal Statistical Society, Series B*, 37:77–87.

Box, G.E.P. and Draper, N.R. (1959). A basis for the selection of a response surface design. *Journal of the American Statistical Association*, 54:622–654.

Box, G.E.P. and Draper, N.R. (1975). Robust designs. *Biometrika*, 62:347–352.

Box, G.E.P. and Draper, N.R. (1987). *Empirical Model-Building and Response Surfaces*. John Wiley & Sons, New York.

Box, G.E.P. and Wilson, K.B. (1951). On the experimental attainment of optimum conditions (with discussion). *Journal of the Royal Statistical Society, Series B*, 13:1–45.

Cambanis, S. (1985). Sampling design for time series. In Hannan, E.J., Krishnaiah, P.R., and Rao, M.M., editors, *Handbook of Statistics 5*. Elsevier Science Publishers, 337–362.

Chang, Y-J. and Notz, W.I. (1996). Model robust design. In Gosh, S. and Rao, C.R., editors, *Handbook of Statistics*, volume 13. Elsevier, Amsterdam, 1055–1098.

Cheng, M., Hall, P. and Titterington, D.M. (1998). Optimal designs for curve estimation by local linear smoothing. *Bernoulli*, 4(1):3–14.

Cleveland, W.S. and McRae, J.E. (1988). The use of loess and STL in the analysis of atmospheric CO_2 and related data. Technical Report 67, AT&T Bell Laboratories.

von Doderer, H. (1951). *Die Strudlhofstiege*. Biederstein Verlag, München.

Evans, J.W. and Manson, A.R. (1978). Optimum experimental designs in two dimensions using minimum bias estimation. *Journal of the American Statistical Association*, 73:171–176.

Fang, K.-T. and Wang, Y. (1994). *Number-theoretic Methods in Statistics*. Chapman & Hall, London.

Fang, K.-T. (1980). The uniform design: application of number theoretic methods in experimental design. *Acta Math. Appl. Sinica*, 3:363–372.

Fedorov, V.V. and Hackl, P. (1994). Optimal experimental design: Spatial sampling. *Calcutta Statistical Association Bulletin*, 44:57–81.

Fedorov, V.V. and Khabarov, V. (1986). Duality of optimal design for model discrimination and parameter estimation. *Biometrika*, 73(1):183–190.

Fedorov, V.V., Montepiedra, G. and Nachtsheim, C.J. (1999). Design of experiments for locally weighted regression. *Journal of Statistical Planning and Inference*, 81:363–382.

Fedorov, V.V. (1989). Kriging and other estimators of spatial field characteristics. *Atmospheric Environment*, 23(1):175–184.

Felsenstein, K. (1992). Optimal Bayesian design for discriminating among rival models. *Computational Statistics and Data Analysis*, 14:427–436.

Herzberg, A. and Huda, S. (1981). A comparison of equally spaced designs with different correlation structures in one and more dimensions. *The Canadian Journal of Statistics*, 9:203–208.

John, P.W.M., Johnson, M.E., Moore, L.M. and Ylvisaker, D. (1995). Minimax distance designs in two-level factorial experiments. *Journal of Statistical Planning and Inference*, 44:249–263.

Johnson, M.E., Moore, L.M. and Ylvisaker, D. (1990). Minimax and maximin distance designs. *Journal of Statistical Planning and Inference*, 26:131–148.

Koehler, J.R. and Owen, A.B. (1996). Computer experiments. In Gosh, S. and Rao, C.R., editors, *Handbook of Statistics*, volume 13. Elsevier, Amsterdam, 261–308.

Matern, B. (1960). *Spatial Variation*, reprint. Springer Verlag, Heidelberg.

McArthur, R.D. (1987). An evaluation of sample designs for estimating a locally concentrated pollutant. *Communications in Statistics — Simulation*, 16(3):735–759.

Morris, M.D. and Mitchell, T.J. (1995). Exploratory designs for computational experiments. *Journal of Statistical Planning and Inference*, 43:381–402.

Müller, H.G. (1984). Optimal designs for nonparametric kernel regression. *Statistics & Probability Letters*, 2:285–290.

Müller, H.G. (1988). *Nonparametric Regression Analysis of Longitudinal Data*, volume 46 of *Lecture Notes in Statistics*. Springer—Verlag.

Müller, W.G. (1992). Optimization of a monitoring network: The moving local regression technique. In Pázman, A. and Volaufová, J., editors, *Proceedings of PROBASTAT '91*. Printing-house of the Technical University Liptovsky Mikulas, 116–122.

Müller, W.G. (1996). Optimal designs for moving local regression. *Journal of Statistical Planning and Inference*, 55:389–397.

Müller, W.G. (2000). Coffee-house designs. In Atkinson, A.C., Bogacka, B., and Zhigljavsky, A.A., editors, *Optimum Design 2000*. Kluwer, 241–248.

Müller, W.G. and Ponce de Leon, A.C.M. (1996). Discriminating between two binary data models: sequentially designed experiments. *Journal of Statistical Computation and Simulation*, 55:87–100.

Nychka, D., Yang, Q. and Royle, J.A. (1996) Constructing spatial designs using regression subset selection. In Barnett, V. and Turkman, K.F., editors, *Statistics for the Environment 3: Sampling and the Environment*, Wiley, New York, 131–154.

O'Hagan, A. (1978). Curve fitting and optimal design for prediction. *Journal of the Royal Statistical Society, Series B*, 40(1):1–42.

Park, S.H. (1978). Experimental designs for fitting segmented polynomial regression models. *Technometrics*, 20:151–154.

Pettitt, A.N. and McBratney, A.B. (1993). Sampling designs for estimating spatial variance components. *Applied Statistics*, 42(1):185–209.

Ponce de Leon, A.C. and Atkinson, A.C. (1991). Optimum experimental design for discriminating between two rival models in the presence of prior information. *Biometrika*, 78:601–618.

Pukelsheim, F. and Rosenberger, J.L. (1993). Experimental design for model discrimination. *Journal of the American Statistical Association*, 88(422):642–649.

Riccomagno, E., Schwabe, R. and Wynn, H.P. (1997). Lattice-based optimum designs for fourier regression. *The Annals of Statistics*, 25(6):2313–2327.

Royle, J.A. and Nychka, D. (1998) An algorithm for the construction of spatial coverage designs with implementation in SPLUS. *Computers & Geosciences*, 24(5):479–488.

Tobias, R. (1995). *SAS QC Software, Vol. 2: Usage and Reference.* SAS Institute, Cary, NC.

Wahba, G. (1992). *Spline Models for Observational Data.* SIAM, Philadelphia.

Yfantis, E.A., Flatman, G.T. and Behar, J.V. (1987). Efficiency of kriging estimation for square, triangular, and hexagonal grids. *Mathematical Geology*, 19:183–205.

Ylvisaker, D. (1975). Design on random fields. *A Survey on Statistical Design and Linear Models*, 593–607.

5
Designs for Spatial Trend Estimation

*" Come up, my dear sir," said Holmes's voice from above.
"I hope you have no designs upon us such a night as this."*
Conan-Doyle (1904)

Spatial data collection schemes usually exhibit two decisive features that distinguish them from classical regression designs (cf. the review paper by Fedorov (1996)). First, spatial observations are often determined by local correlations, which are unaccounted for by standard o.d.e. techniques as presented in Chapter 3. Second, at least in the framework of our basic random field model (2.1), there is implicitly no possibility for instantaneously replicated measurements, so that the classical concept of design measures is not applicable. This chapter gives an extension of information matrices with a different definition of design measures, that will allow them to be used for constructing design optimization algorithms for the estimation of a spatial trend parameter β.

For this purpose we require the (standardized) information matrix corresponding to the best linear unbiased estimate $\hat{\beta}$, cf. (2.3), of a random field, which can be written as (see e.g. Näther (1985a))

$$M(A) = \frac{1}{N_A} \sum_{x \in A} \sum_{x' \in A} \dot{\eta}(x) [C^-(A)]_{x,x'} \dot{\eta}^T(x'). \qquad (5.1)$$

Here the finite subset $A = \{x_i \neq x_{i'}; i \neq i' = 1, \ldots, n_A \equiv N_A\}$ with $p_1 = p_2 = \ldots = p_{n_A} = \frac{1}{n_A}$ forms the so-called replication-free design ξ_{n_A}. We also use now the notation $[C(A)]_{ii'} = c(x_i, x_{i'}; \theta)$ to emphasize the dependence of C on the design and we assume knowledge of θ throughout this chapter. Efficient designs for the estimation of θ (and both β and θ) are presented in Chapter 6. The optimum n-point design (for estimating β) is a solution of the maximization problem

$$\max_{A \subset \mathcal{X}, n_A \leq n} \Phi[M(A)]. \tag{5.2}$$

Unfortunately the information matrix (5.1) does not share all desirable properties of its analogue (3.2) in the standard regression setting, which allow for the construction of design algorithms similar to those in Section 3.3. For instance, it is neither convex nor additive in x and thus information from different design points can no longer be separated as nicely as before. Furthermore, as already mentioned, due to the fact that instantaneous replications are not possible, the classical definition and interpretation of design measures as the proportion of replications (see Kiefer (1959) and Chapter 3) is generally not useful. There are, however, still some design measures $\xi_A(\cdot)$, which have a good interpretation, namely those which are uniformly distributed over the set A:

$$\xi_A(x) = \frac{\mathcal{I}_A(x)}{n_A},$$

where $\mathcal{I}_A$ is the indicator of A. For the construction of design optimization algorithms similar to those in Section 3.3 it will be necessary to allow for any other design measure from Ξ.

Let us now introduce the ratio $\frac{\xi(x)}{\xi_{\max}}$, where $\xi_{\max} = \max\{\xi(x) : x \in \mathcal{X}\}$ for norming purposes. Our aim is to suppress part of the signal $\eta(x, \beta)$ of the random field if this ratio is small, which means to give relatively more weight to the signal where the ratio is one.

5.1 Approximate Information Matrices

Note, that for any design A with associated uniform design measure ξ_A

$$\frac{\xi_A(x)}{\xi_{\max}} = 1; \quad x \in A$$

5.1 Approximate Information Matrices

and we have a respective information matrix $M(A)$ from (3.2). Other design measures $\xi(\cdot)$ do not correspond to real experiments, but it is still important to associate an information matrix to each $\xi(x)$ to allow for computational methods that are based upon continuously defined measures. Such information matrices should be extensions of $M(A)$, in the sense that $M(A)$ is retained for all measures $\xi_A(\cdot)$ that correspond to replication-free designs. One such extension is provided by $M(S_\xi)$, the total information matrix of the support of a measure $\xi(\cdot)$, see the arguments given in Appendix A.5.1. We can now further, for instance, employ parameterized approximations of these extensions of the form

$$M^{(\epsilon)}(\xi) = f\left(\frac{\xi(x)}{\xi_{\max}}, \dot{\eta}(\cdot), C(\cdot), \epsilon\right).$$

In such an approximation it is desirable to retain the property that small values of $\xi(x)$ mean that little information from x is contributed to $M^{(\epsilon)}(\xi)$, and larger values of $\xi(x)$ the opposite, which is evident for classical design measures. Also for constructing gradient type design algorithms it should be guaranteed that a subsequently applied design criterion remains a continuous and differentiable, and possibly concave function of the design. Two such parameterized approximations of extensions of the information matrix have been proposed by Müller and Pázman (1998) and Pázman and Müller (1998) respectively:

$$M_1^{(\epsilon)}(\xi) = \sum_{x,x' \in S_\xi} \left(\frac{\xi(x)}{\xi^{(\epsilon)}}\right)^\epsilon \dot{\eta}(x)[C^-(S_\xi)]_{x,x'}\dot{\eta}^T(x')\left(\frac{\xi(x')}{\xi^{(\epsilon)}}\right)^\epsilon, \quad (5.3)$$

and

$$M_2^{(\epsilon)}(\xi) = \sum_{x,x' \in S_\xi} \dot{\eta}(x)[C(S_\xi) + W^{(\epsilon)}(\xi)]_{x,x'}^{-1}\dot{\eta}^T(x'), \quad (5.4)$$

where $W^{(\epsilon)}(\cdot)$ is a diagonal matrix with entries

$$[W^{(\epsilon)}(\xi)]_{x,x} = \ln\left(\frac{\xi^{(\epsilon)}}{\xi(x)}\right)^\epsilon.$$

In both choices $\epsilon > 0$ is a small number, and $\xi^{(\epsilon)} = \left(\sum_{x \in \mathcal{X}} \xi^{\frac{1}{\epsilon}}(x)\right)^\epsilon$ is a continuous approximation of $\xi_{\max}$ with the limit $\lim_{\epsilon \to 0} \xi^{(\epsilon)} = \xi_{\max}$. Evidently both matrices $M_1^{(\epsilon)}(\xi_A)$ and $M_2^{(\epsilon)}(\xi_A)$ coincide with $M(A)$

on replication-free designs $\xi = \xi_A$ for $\epsilon = 0$ and thus order them in the same way.

The motivation for the construction of (5.3) is the following: If in the definition of $M_1^{(\epsilon)}(\xi)$ we set $\tilde{\eta}(x) = \left(\frac{\xi(x)}{\xi_{\max}}\right)^\epsilon \dot{\eta}(x)$ we obtain the total information matrix with $\tilde{\eta}(x)$ in place of $\dot{\eta}(x)$. But the expression $|\tilde{\eta}^T(x)\beta|$ is smaller than $|\dot{\eta}^T(x)\beta|$ (and thus $|\eta(x,\beta)|$) for every $\beta \in \mathbb{R}^m$ (except for an x where $\xi(x) = \xi_{\max}$). So, if $\xi(x) < \xi_{\max}$ the signal part (i.e. the trend) of the observation $y(x)$ is suppressed by using $\tilde{\eta}(x)$ instead of $\dot{\eta}(x)$. A large $\xi(x)$ means a slight suppression, $\xi(x) = 0$ means total suppression, $\xi(x) = \xi_{\max}$ means no suppression. In other words, there is no contribution to the information matrix when $\xi(x) = 0$ and there is only a fraction of the information when $\xi(x) < \xi_{\max}$.

The motivation for (5.4) is similar. Suppose that to the existing noise $\varepsilon(x)$ some supplementary design depending (virtual) noise $\varepsilon^+(x)$ is added. The smaller $\frac{\xi(x)}{\xi_{\max}}$, the larger has to be the variance of $\varepsilon^+(x)$. So let the mean and variance of $\varepsilon^+(x)$ be $E[\varepsilon^+(x)] = 0$, $\mathrm{Var}[\varepsilon^+(x)] = \ln\left(\frac{\xi_{\max}}{\xi(x)}\right)^\epsilon$, respectively. If $\xi(x) = \xi_{\max}$ then $\mathrm{Var}[\varepsilon^+(x)] = 0$, hence also $\varepsilon^+(x) = 0$ and at such x there is no additional noise. If $\xi(x) \to 0$, then $\mathrm{Var}[\varepsilon^+(x)] \to \infty$, hence there is no information from the observation at such an x. In the cases which are between, i.e. $0 < \xi(x) < \xi_{\max}$, the additional noise $\varepsilon^+(x)$ suppresses only part of the contribution to the information matrix from an observation at x. The matrix $W^{(\epsilon)}(\cdot)$ serves as a continuous approximation of the covariance matrix of this supplementary noise.

Note that in the given context the design measure ξ is a theoretical construction which does not correspond to a real experiment, except for $\xi = \xi_A$. This also leads to a new interpretation of the design measure in the approach presented here. While in the classical theory $\xi(x)$ defines the number of replications of observations at x, here $\xi(x)$ defines the amount of suppression of the informative component of the observation at x.

It is remarkable that for both choices (5.3) and (5.4) of the approximate information matrix we have

$$\lim_{\epsilon \to 0} n_{S_\xi}^{-1} M^{(\epsilon)}(\xi) = M(S_\xi), \tag{5.5}$$

which is a concave function of ξ. Moreover, the mapping of ξ into $\Phi[M(S_\xi)]$ is concave for all concave criteria, such as those presented in Section 3.2 (for further justification of this mapping see Appendix

A.5.1). This property of $\Phi[\cdot]$ is a good justification of using a small ϵ and the limit $\epsilon = 0$ in the definition of the two information matrices (and their derivatives) to construct numerical algorithms for finding optimum designs (cf. Müller and Pázman (1998) and Müller and Pázman (1999)).

However, there is a basic restriction in exploiting the concavity property. If we maximize $\Phi[M(S_\xi)]$ over ξ we obtain that the optimum design is supported on the whole $\mathcal{X}$ which is a useless solution. We have to introduce an a priori restriction on the number of design points, i.e. we have to maximize $\Phi[M^{(\epsilon)}(\xi)]$ over the set

$$\Xi_n = \{\xi \in \Xi : n_{S_\xi} \leq n\},$$

which is nonconvex (see Appendix A.5.2). This restriction destroys the concavity properties of the criterion function and a formal equivalence theorem in the usual sense and thus a simple rule for determining whether a given design is optimum cannot be found. Nevertheless, the introduction of design measures allows us to change the original discrete optimization problem into a continuous one. This enables us to construct standard (gradient) algorithms to find local minima of the criterion functions as will be done in the following sections.

5.2 Replication-Free Designs

We will discuss here the construction of designs for the replication-free uncorrelated case, i.e. $C(A) = I_{n_A}$, as a special case of the above framework. Investigations of replication-free designs have been initiated by Wynn (1982) and Cook and Fedorov (1995) embed Wynn's work into the setting of constrained optimum designs.

Since here, there is no hope of finding a (global) equivalence theorem by the principle of nonconcavity (as is the case for the exact design problem in general, see Pukelsheim (1993)), it is extremely useful to have several parallel methods for finding an optimum design. It turns out that the approximate information matrix $M_1^{(\epsilon)}(\cdot)$ provides a good basis for the construction of an efficient competitive algorithm in the uncorrelated case (cf. Müller and Pázman (1998)).

5. Designs for Spatial Trend Estimation

An Approach Based on Approximate Information Matrices

Let us now follow the spirit of Chapter 3 and apply a differentiable optimality criterion $\Phi[\cdot]$ (see section 3.3) to the approximate information matrix $M_1^{(\epsilon)}(\xi)$. Take fixed design measures ξ and ξ', and consider the directional derivative of $\Phi[\cdot]$ at the point ξ in the direction to ξ', which is equal to

$$\phi^{(\epsilon)}(\xi, \xi') = \lim_{\alpha \to 0} \frac{\partial \Phi[M_1^{(\epsilon)}(\xi_\alpha)]}{\partial \alpha},$$

where

$$\xi_\alpha(x) = (1-\alpha)\xi(x) + \alpha \xi'(x),$$

and $\alpha \in [0,1)$. The expression $\phi^{(\epsilon)}(\cdot,\cdot)$ will eventually be used as a kind of sensitivity function. Actually, we are interested in computing $\phi^{(\epsilon)}(\cdot,\cdot)$ for a small ϵ, since in the limit both matrices $M_1^{(\epsilon)}(\xi)$ (and $M_2^{(\epsilon)}(\xi)$) approach the basic, but discontinuous extension $M(S_\xi)$ of the information matrix. Computing the directional derivative of $\Phi[M_1^{(\epsilon)}(\xi)]$ for a small ϵ it becomes clear that its main term is just the limit for $\epsilon \to 0$ and we take this limit as the basis of an algorithm. Following these arguments leads us to consider (see Appendix A.5.3)

$$\lim_{\epsilon \to 0} \frac{1}{\epsilon} \phi^{(\epsilon)}(\xi, \xi') = \operatorname{tr}\left\{ \nabla \Phi[M_1(S_\xi)] \lim_{\epsilon \to 0} \lim_{\alpha \to 0} \frac{1}{\epsilon} \frac{\partial M_1^{(\epsilon)}(\xi_\alpha)}{\partial \alpha} \right\} \quad (5.6)$$

as the approximate sensitivity function.

A Directional Derivative Based Algorithm

Note that unlike the classical case there is no limitation of support points. Thus, as noted before, a design that observes the whole process (i.e. at every point of the grid $\mathcal{X}$, such that $\dot{\eta}(x) \neq 0$) is always an optimum design. Hence, we not only have to care about maximizing the criterion but also about limiting the necessary number of observations. So we can maximize $\Phi[M_1^{(\epsilon)}(\xi)]$ under the condition that n_{S_ξ}, the number of points in the support of S_ξ equals n, e.g. by applying the Lagrange multiplier method (with the multiplier denoted by κ)

$$\max_{\xi \in \Xi_n} \Phi[M_1^{(\epsilon)}(\xi)] \simeq \max_{\xi \in \Xi, \kappa} \left\{ \Phi[M_1^{(\epsilon)}(\xi)] + \kappa \ln\left(\frac{n_\xi^{(\epsilon)}}{n}\right) \right\}.$$

Since n_{S_ξ} is a nondifferentiable function, it is approximated by $n_\xi^{(\epsilon)} = \sum_{x \in \mathcal{X}} \xi(x)^\epsilon$, which makes sense due to $\lim_{\epsilon \to 0} n_\xi^{(\epsilon)} = n_{S_\xi}$. Thus the directional derivative (5.6) of $\Phi[M^{(\epsilon)}(\xi)]$ has to be supplemented by (see Appendix A.5.3)

$$\lim_{\epsilon \to 0} \lim_{\alpha \to 0} \frac{1}{\epsilon} \frac{\partial \ln n_{\xi_\alpha}^{(\epsilon)}}{\partial \alpha} = \left[\frac{1}{n_{S_\xi}} \sum_{x \in S_\xi} \frac{\xi'(x)}{\xi(x)} \right] - 1.$$

However, the rate of convergence of this supplement turned out to be sometimes insufficient in the case of very small numbers of observations. For this situation Müller and Pázman (1998) suggest an improved approximation.

Note that although the information matrix $M_1^{(\epsilon)}(\xi)$ is nonlinear in ξ, the limiting directional derivative is linear in the correcting design ξ' (as seen from (A.17) in Appendix A.5.3). This justifies the use of one-point correction algorithms of type (3.11). Therefore the final approximate sensitivity function for a one point design $\xi'(x)$ concentrated at $x \in S_\xi$ can be written as (see Appendix A.5.3)

$$\tilde{\phi}_1(x, \xi) = \left\{ \frac{p(x)}{\xi(x)} - \frac{\mathcal{I}_{B_\xi}(x)}{\xi_{\max} n_{B_\xi}} q(\xi) \right\} + \kappa \left\{ \frac{1}{\xi(x) n_{S_\xi}} - 1 \right\}, \quad (5.7)$$

where

$$q(\xi) = \mathrm{tr} \nabla \Phi[M(S_\xi)] M(S_\xi),$$
$$r(x) = \dot{\eta}^T(x) \nabla \Phi[M(S_\xi)] \dot{\eta}(x),$$

and $B_\xi = \{x \in \mathcal{X} : \xi(x) = \xi_{\max}\}$. Note the similar structure of the approximate and the classical sensitivity function $\phi(x, \xi)$, e.g. for D-optimality we have $\nabla \ln |M| = M^{-1}$, hence $q(\xi) = m$.

Expression (5.7) can be used in place of $\phi(x, \xi)$ in any algorithm of type (3.11). In such an iterative algorithm we will start by using a small $\kappa > 0$. In the case that $S_\xi = A$ is the support of the design it follows from the definition (5.3) that for a small $\epsilon > 0$ we have $M_1^{(\epsilon)}(\xi) \leq M_1^{(\epsilon)}(\xi_A)$ in the Loewner ordering. Hence for increasing criteria functions $\Phi[\cdot]$ (like those given in Section 3.2) one has

$$\Phi[M_1^{(\epsilon)}(\xi_A)] \geq \Phi[M_1^{(\epsilon)}(\xi)].$$

It follows that the iterative computation will stabilize its position at a design with support S_ξ (from $\epsilon = 0$) very close to ξ_A, for some A.

We take this A as the exact design resulting from the computation. If n_A is larger than n, we restart the algorithm using a larger κ and so forth until the restriction is satisfied. Since the optimized function is nonconcave by principle (see the earlier remark) we may obtain a local minimum. The procedure should thus be restarted for various starting designs, and the different resulting designs should be compared according to the criterion values.

As an example consider D-optimum design for a linear regression on the unit-square, i.e. we assume $\dot{\eta}^T(x) = (1, x_1, x_2)$; $-1 \leq x_1, x_2 \leq 1$; $\Phi[M] = \ln|M|$. A 20×20 point grid was used to approximate the design space. On this grid only 100 observations (without replications) were allowed. The supporting points of the resulting design are all points outside a circle as given in Figure 5.1. This design was constructed by a simple one-point correction algorithm (with $\alpha^{(s)} = \frac{1}{n+s}$) based upon $\tilde{\phi}_1(x, \xi)$ (by increasing κ up to the value 0.033) after only 200 iterations.

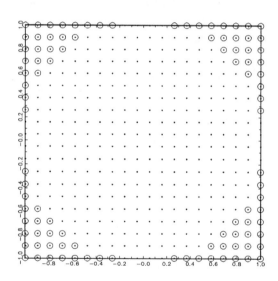

FIGURE 5.1. A replication-free optimum design for linear regression

Alternative Approaches

If the discretization $\bar{\mathcal{X}}$ of the design region contains $\mathcal{N}$ points, it is evident that we can in principle find a replication-free n_A point design A by enumerating all $\binom{\mathcal{N}}{n_A}$ possible subsets and select the one with the maximum criterion value $\Phi[M(A)]$. An efficient branch and bound algorithm for that purpose is given by Rasch et al. (1997). In their example 1 for instance, they employ this algorithm to find a 9-point replication-free D-optimum design for the setting

$$\dot{\eta}^T(x) = (1, e^{-0.2x}, -1.4xe^{-0.2x}), \qquad x \in \{0, 1, \ldots, 25\}.$$

However, by the approximate algorithm given above the same optimum design $\xi_D^* = \{0, 1, 4, 5, 6, 22, 23, 24, 25\}$ is calculated from only a few iterations.

Moreover, if $\mathcal{N}$ (and n_A) is large (as is frequently the case in spatial problems) the computational burden may become prohibitive and approximative solutions must be employed. An alternative algorithm that can be applied for the numerical construction of optimum designs for uncorrelated observations without replications was given by Fedorov (1989). It is based on the fact that optimum replication-free designs can be considered as special solutions of the more general problem of optimum designs with constrained measures, i.e.

$$A^* = S_{\xi_{n_A}^*}, \qquad \text{with} \quad \xi_{n_A}^* = \arg\max_{\xi \in \Xi_\omega} \Phi[M(\xi)],$$

where $\Xi_\omega = \{\xi : \xi(x) \leq \omega(x); x \in \mathcal{X}\}$ and the particular choice $\omega(\cdot) = \frac{1}{n_A}$. By approximating the discrete probability measures by continuous ones Wynn (1982) finds (cf. also Fedorov (1996)) that

$$\min_{x \in S_{\xi^*}} \phi(x, \xi^*) \geq \max_{x \in \mathcal{X} \setminus S_{\xi^*}} \phi(x, \xi^*) \qquad (5.8)$$

is a necessary and sufficient condition for $\xi^* \in \Xi_\omega$ to be optimum with respect to the criterion $\Phi[\cdot]$. Note, that (5.8) is not an equivalence theorem for exact designs, since it only holds in the space of designs with potentially non-countable support.

However, it indicates that reasonable algorithms must attempt to create two sets of points (S_{ξ^*} and its complement) that come close to obeying condition (5.8). The basic version of such an algorithm has a very similar iterative structure to the standard algorithms presented in section 3.2. It consists in a simple exchange of points from the two sets S_{ξ_s} and $\bar{\mathcal{X}}_s \setminus S_{\xi_s}$ at every iteration s, namely

$$\xi_{s+1} = \left\{ \xi_s \setminus \left\{ x_s^-, \frac{1}{n_s} \right\} \right\} \cup \left\{ x_s^+, \frac{1}{n_s} \right\},$$

where

$$x_s^+ = \arg\max_{x \in \bar{\mathcal{X}}_s \setminus S_{\xi_s}} \phi(x, \xi_s) \quad \text{and} \quad x_s^- = \arg\min_{x \in S_{\xi_s}} \phi(x, \xi_s).$$

The set $\bar{\mathcal{X}}_s$ now carries a subscript to indicate the possibility of changing the discretization of $\mathcal{X}$ at every iteration. A generalization of this algorithm to exchanges of more than one point at every step is straightforward.

The algorithm was employed to once more calculate the design in Figure 5.1, which thus may truly be the correct exact optimum design. It is hard to say, which of the two methods converges more quickly. However, since there exists no equivalence theorem for exact designs by principle of nonconcavity of the optimization problem (as discussed in the previous section or Pukelsheim (1993), page 305) it is here of particular importance to have several methods for the computation of the same problem.

It is important to note that for the convergence of the algorithm to eventual fulfillment of (5.8) it is necessary, that the distance between neighboring points in $\bar{\mathcal{X}}_s$ tends to zero. A proof of this convergence and variants of the algorithm can be found in Fedorov (1989). For practical needs it is usually sufficient to use a reasonably dense grid. Rasch et al. (1997) compare algorithms of the above type to enumeration algorithms for some simple examples and their reported efficiencies of the former are remarkable.

5.3 Designs for Correlated Fields

Employing approximate information matrices for the construction of replication-free design for the estimation of the parameter β presented in the previous section gives a basis for applying the same idea in the more general correlated observation case, characterized by information matrices of the general form (5.1). However, that this might be not so straightforward can be seen by reconsidering the role of the design measure in matrix $M_1^{(\epsilon)}$ in the correlated setup. The random field $y(x)$ contains two components. The trend (or signal) $\eta(x, \beta)$, which is of interest, and the nuisance component, which is the correlated random noise $\varepsilon(x) = y(x) - \eta(x, \beta)$. We see from (5.4) that by taking $\xi(x) < \xi_{\max}$ we suppress the influence of $\dot{\eta}(x)$ in the expression for $M_1^{(\epsilon)}(\xi)$. This suppression of the signal component works well when the correlations between observations $y(x)$ at different points $x \in \mathcal{X}$ are zero or small. However, if observations at different points are highly correlated, then observations at points with even completely suppressed signal still provide some information through the noise at other points and therefore contribute indirectly to the increase in information about the parameter β. Thus, the approach from the previous section can only be used for uncorrelated or slightly correlated observations (see Müller and Pázman (1998)).

Fortunately, this is not the case for the approximate information matrix $M_2^{(\epsilon)}$ (see Pázman and Müller (1998)), which in contrast to $M_1^{(\epsilon)}$, was constructed by adding a supplementary white noise depending upon the design and a parameter ϵ. If at a specific site $\xi(x) \to 0$, then $\text{Var}[\varepsilon^+(x)] \to \infty$, hence $\eta(x, \beta)$ but also $\varepsilon(x)$ is totally dominated by $\varepsilon^+(x)$ and there is no information at all from the observation at x. Thus, there is also no more information transfer to other observations due to the correlation structure and the approximation should work for arbitrary correlation schemes. A second property of (5.4) is that again uniform design measures correspond then to exact designs, and in this case the approximate information matrix also coincides with the standard one.

Moreover, Pázman and Müller (2000) succeed to show that the problem of finding

$$\xi^* = \arg\max_{\xi \in \Xi_n} \Phi[M_2^{(\epsilon)}(\xi)] \qquad (5.9)$$

is completely analogous to solving (5.2) in the sense that S_{ξ^*} is a solution to it (cf. Appendix A.5.7). Thus, it is still possible to use the so attractive concept of (approximate) design measures similar to the classical (uncorrelated) theory, which was considered as unfeasible. E.g. Näther (1985a) on p. 84 argues that "... the corresponding information matrix [...] allows no generalization in the direction of design measures, at least not in the same manner as known from Kiefer's theory".

An Approach Based on Approximate Information Matrices

Without restrictions on the number of observations the uniform design on $A = \bar{\mathcal{X}}$ is again always an optimum design, (i.e. observations should be at each point of $\bar{\mathcal{X}}$). But meaningful designs ξ are the ones with the number of support points n_{S_ξ} restricted to a given n. A standard approach is to use Lagrangian optimization (c.f. Section 5.2), but the respective multiplier has the practical disadvantage that it cannot be chosen in advance, and is to be determined empirically during the computations.

Therefore, for the algorithm in this section an alternative approach on how to restrict the number of observations will be employed. The idea is to modify the definition of the approximate information matrix (5.4) by making the additional noise $\varepsilon^+(x)$ dependent upon n, the upper bound on the number of design points. Let us take $\kappa = \frac{1}{n}$ and cut the design measure ξ on the level κ, i.e. let us introduce

$$\xi^{\#}(x) = \max\{0, \xi(x) - \kappa\}$$

and

$$\xi^{\#}_{\max} = \max\{\xi^{\#}(x); x \in \mathcal{X}\}.$$

The approximate information matrix (5.4) is modified by inserting $\xi^{\#}_{\max}$ and $\xi^{\#}(x)$ instead of $\xi_{\max}$ and $\xi(x)$ in the definition of the variance of the additional noise $\varepsilon^+(x)$, given as

$$\text{Var}[\varepsilon^+(x)] = \ln\left(\frac{\xi^{\#}_{\max}}{\xi^{\#}(x)}\right)^{\epsilon}.$$

So if $\xi_{\max} > \kappa$, observations from points $x \in \mathcal{X}$ with $\xi(x) < \kappa$ give no information at all, since $\xi^{\#}(x) = 0$, and $\text{Var}[\varepsilon^+(x)] = \infty$. In the unrestricted case of the previous section this is achieved only if $\xi(x) = 0$.

To make the approximate information matrix differentiable we require the expressions $_\epsilon\xi(x)$ and $_\epsilon\xi$ in (5.10), which are smooth approximations of $\xi^{\#}(x)$, and $\xi^{\#}_{\max}$ respectively, such that $\lim_{\epsilon \to 0} {_\epsilon\xi}(x) = \xi^{\#}(x)$, and $\lim_{\epsilon \to 0} {_\epsilon\xi} = \xi^{\#}_{\max}$. Now we can once more (re)define matrix (5.4) as

$$M^{(\epsilon)}_{2,\kappa}(\xi) = \sum_{x,x' \in S_\xi} \dot{\eta}(x)[C(S_\xi) + W^{(\epsilon)}_\kappa(\xi)]^{-1}_{x,x'}\dot{\eta}^T(x'),$$

where $W^{(\epsilon)}_\kappa(\xi)$ is a diagonal matrix with entries

$$[W^{(\epsilon)}_\kappa(\xi)]_{x,x} = \ln\left(\frac{_\epsilon\xi}{_\epsilon\xi(x)}\right)^{\epsilon}. \tag{5.10}$$

Specifically, one may choose

$$_\epsilon\xi(x) = [\kappa^{\frac{1}{\epsilon}} + \xi^{\frac{1}{\epsilon}}(x)]^{\epsilon} - \kappa, \tag{5.11}$$

and
$$_\epsilon\xi = [\kappa^{\frac{1}{\epsilon}} + \sum_{x\in\bar{\mathcal{X}}}\xi^{\frac{1}{\epsilon}}(x)]^\epsilon - \kappa. \qquad (5.12)$$

Note that the matrix $M_{2,\kappa}^{(\epsilon)}(\xi)$ is a direct extension of $M_2^{(\epsilon)}(\xi)$, because when $\kappa = 0$, we have for every $\epsilon > 0$

$$M_{2,0}^{(\epsilon)}(\xi) = \sum_{x,x'\in S_\xi} \dot{\eta}(x)[C(S_\xi) + W^{(\epsilon)}(\xi)]_{x,x'}^{-1}\dot{\eta}^T(x'),$$

which coincides with (5.4). We also see that for $\xi(x) \geq \kappa$ for every $x \in S_\xi$,

$$\lim_{\epsilon\to 0} M_{2,\kappa}^{(\epsilon)}(\xi) = \sum_{x,x'\in S_\xi} \dot{\eta}(x)C^{-1}(S_\xi)\dot{\eta}^T(x'),$$

which is the same as $\lim_{\epsilon\to 0} M_2^{(\epsilon)}(\xi)$ (see (5.5)). So in this case κ has no influence on the limit value of the information matrix.

On the other hand, if $\xi(x) < \kappa$ for some $x \in S_\xi$, we see that an additional noise $\varepsilon^+(\cdot)$ remains also in the limit $\epsilon \to 0$ (according to Appendix A.5.4 its variance is $\frac{\min(\kappa,\xi_{\max})}{\xi(x)}$). So points with $\xi(x) > \kappa$ are preferred, but their number is restricted by $n = \frac{1}{\kappa}$. Thus κ can be effectively used to control the number of support points of the resulting optimum design in advance.

A Directional Derivative Based Algorithm

On the basis of the directional derivatives of $\Phi[M_{2,\kappa}(\xi)]$ again a one-point correction algorithm for design optimization can be built; other algorithms analogous to those known from classical design theory are possible.

The practical application of such an algorithm consists of the following sequence of steps:

- **starting conditions**

    We shall start the algorithm with a design supported on the whole set $\bar{\mathcal{X}}$. Usually we take as a starting design either the uniform design $\xi_{\bar{\mathcal{X}}}(x) = \frac{1}{n_{\bar{\mathcal{X}}}}$, or a design which is close to the supposed optimum design but supported on the whole set $\bar{\mathcal{X}}$.

    We choose $\kappa = \frac{1}{n}$ (or a slightly larger κ to avoid the influence of numerical errors), where n is the intended number of points in the design.

- **the s-th step**

  The iterations consist of stepwise one-point corrections of the design as in (3.11). That means that at the s-th step we increase the measure at the point $x^{(s)}$, where the directional derivative in the direction of $\delta_{x,x^{(s)}}$ is maximized, i.e.

  $$x^{(s)} = \arg\max_{x \in \tilde{\mathcal{X}}_s} \tilde{\phi}_2(x, \xi_{(s)}), \qquad (5.13)$$

  where $\tilde{\phi}_2$ is defined similarly as (5.7) - see below and Appendix A.5.4. Unfortunately unlike (5.7) it cannot be given in closed form and thus a detailed description of one iteration step is given in the next subsection.

- **stopping considerations**

  Let ξ^+ be the resulting design after a reasonably large number of iterations. If ξ^+ is not uniform, it is definitely a false optimum and we have to choose another starting design.

  If ξ^+ is uniform, we might be trapped at a local optimum. To check for that we insert a proportionality parameter ρ for re-weighting the effect of the additional noise into the definition of the approximate information matrix (for details see the next subsection). Then the iterations for a larger $\rho > 1$ and some different starting design (possibly ξ^+) are restarted. One has to be careful however in increasing ρ since eventually the resulting design may collapse to a nonsensical solution.

  When a design $\xi^{++} \neq \xi^+$ is obtained, it means that ξ^+ was only a local maximum of the criterion function. Since this function depends upon the choice of ρ, its local maxima should be shifted by the change of ρ, which explains why $\xi^+ \neq \xi^{++}$. Thus, we have to increase ρ once more.

  On the other hand the result $\xi^+ = \xi^{++}$ indicates that ξ^+ gives the absolute maximum, i.e. the design ξ^+ is optimum as required, since by changing ρ we change the matrix $M_{2,\kappa}^{(\epsilon)}(\xi)$, hence the function $\Phi[M_{2,\kappa}^{(\epsilon)}(\xi)]$, but we do not change $\arg\max_\xi \Phi[M_{2,\kappa}^{(\epsilon)}(\xi)]$, and the global maximum does not depend upon ρ.

5.3 Designs for Correlated Fields

For example, in the simple linear regression

$$y(x) = \begin{pmatrix} 1 \\ x_1 \\ x_2 \end{pmatrix}^T \begin{pmatrix} \beta_1 \\ \beta_2 \\ \beta_3 \end{pmatrix} + \varepsilon(x) \qquad x \in [-1,1]^2 \qquad (5.14)$$

with covariance function $c(x, x') =$

$$\begin{cases} (1 - |x_1 - x_1'|)(1 - |x_2 - x_2'|); & \max(|x_1 - x_1'|, |x_2 - x_2'|) < 1 \\ 0; & \max(|x_1 - x_1'|, |x_2 - x_2'|) \geq 1 \end{cases}, \quad (5.15)$$

the algorithm results (after just a few steps) in a D-optimum 9-point design $\xi_{D,9}^* = \{(-1,-1),(-1,0),(-1,1),(0,-1),(0,0),(0,1),(1,-1),(1,0),(1,1)\}$ (see Figure 5.2), which coincides with the well-known D-optimum design for a second order polynomial in the uncorrelated case.

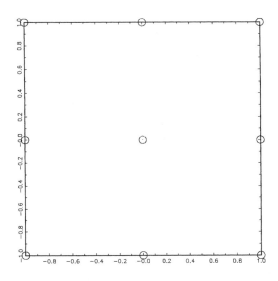

FIGURE 5.2. D-optimum design for a correlated example.

Formal Description of One Iteration Step

Here we give a detailed description how to compute the point $x_{(s)}$ in (5.13) (for a fixed κ and ρ), which is required in the s-th step of the above algorithm. For reasons of notational clarity $x_{(s)}$ is now denoted by either x_r or x_q. Remember that due to (3.11) at each s we have $\xi_{(s)}(x) > 0$ for every $x \in \bar{\mathcal{X}}$, since this is true for $\xi_{(1)}(x)$.

5. Designs for Spatial Trend Estimation

- Define
$$U_\kappa(\xi_{(s)}) = [C(\bar{\mathcal{X}}) + \rho V_\kappa(\xi_{(s)})]^{-1},$$
where $V_\kappa(\xi_{(s)}) = \text{diag}\{\ln \frac{\min\{\xi_{(s),\max},\kappa\}}{\min\{\xi_{(s)}(x),\kappa\}}; x \in \bar{\mathcal{X}}\}$ from Appendix A.5.4.

- Define
$$a(x) = \sum_{z \in \bar{\mathcal{X}}} [U_\kappa(\xi_{(s)})]_{x,z} \dot{\eta}(z),$$

$$g(\xi_{(s)}, x) = a^T(x) \nabla \Phi \left[\sum_{x',z \in \bar{\mathcal{X}}} \dot{\eta}(x') [U_\kappa(\xi_{(s)})]_{x',z} \dot{\eta}^T(z) \right] a(x),$$

see the limit for $\epsilon \to 0$ of (A.19) in Appendix A.5.4.

The 'sensitivity function' $\tilde{\phi}_2$ is defined by the following case distinction:

- Compute
$$q(\xi_{(s)}) = \max_{\{x:\xi_{(s)}(x) \leq \kappa\}} \frac{g(\xi_{(s)}, x)}{\xi_{(s)}(x)}, \tag{5.16}$$

$$x_q = \arg \max_{\{x:\xi_{(s)}(x) \leq \kappa\}} \frac{g(\xi_{(s)}, x)}{\xi_{(s)}(x)}.$$

- Compute (if $\xi_{(s),\max} > \kappa$)
$$r(\xi_{(s)}) = \tag{5.17}$$
$$\max_{\{x:\xi_{(s)}(x) > \kappa\}} \frac{\left\{ g(\xi_{(s)}, x) - \left[\sum_{x' \in \mathcal{X}} g(\xi_{(s)}, x')\right] \mathcal{I}_{B_{\xi_{(s)}}}(x) n_{B_{\xi_{(s)}}}^{-1} \right\}}{\xi_{(s)}(x) - \kappa},$$

$$x_r = \arg \max_{\{x:\xi_{(s)}(x) > \kappa\}} \frac{\left\{ g(\xi_{(s)}, x) - \left[\sum_{x' \in \mathcal{X}} g(\xi_{(s)}, x')\right] \mathcal{I}_{B_{\xi_{(s)}}}(x) n_{B_{\xi_{(s)}}}^{-1} \right\}}{\xi_{(s)}(x) - \kappa}.$$

- At each step:
  - if $q(\xi_{(s)}) < -\epsilon'$ take $x_{(s)} = x_r$,

- if $q(\xi_{(s)}) \in [0 \pm \epsilon']$ and $r(\xi_{(s)}) < 0$ take x_q,
- if $q(\xi_{(s)}) \in [0 \pm \epsilon']$ and $r(\xi_{(s)}) \geq 0$ take x_r,
- if $q(\xi_{(s)}) > \epsilon'$ take x_q.

Here $\epsilon' > 0$ is a fixed small number smoothing the influence of numerical errors on checking the theoretical equation $q(\cdot) = 0$.

Alternative Approaches

It is evident that in much the same way as in the uncorrelated replication-free setting for small scale design problems enumerative methods could be employed for finding an exact solution. Frequently, however, efficient search algorithms are used to identify designs that may be close to the optimum. E.g. Schilling (1992) uses simulated annealing to construct spatial designs for estimating a constant surface model under various correlation structures.

For some restricted cases even a closed analytic solution may be available. As an example consider the simple linear regression, i.e. $\dot{\eta}(x) = \binom{1}{x}$ and

$$c(x, x') = \begin{cases} \sigma^2 & x = x' \\ b\sigma^2 & 0 < |x - x'| \leq d \\ 0 & \text{else} \end{cases},$$

which is a discontinuous function depending upon the parameters $b \in [0, 1]$ and $d \in [0, 2]$ ($x, x' \in \mathcal{X} = [-1, 1]$). The case of D-optimum design for $N = 3$ was extensively studied by Müller (1995). Here three types of optimum designs were identified depending upon the choice of b and d. For $b \in (0.33, 0.5]$, for instance, and $d \in [0, 1]$ the optimum three point design is supported on the two endpoints $x_{1,2} = \pm 1$ of the region plus on an additional grid point with $|x_3| < 1 - d$, such that $|x_3|$ is maximum.

When the problems are more complex, again other (approximative) methods have to be employed. Most of these methods originate in time series sampling problems and unfortunately some of them are not easily generalizable to the spatial setting (see e.g. Yuditsky (1974) and especially Saunders and Eccleston (1992)). Also the approach by Spokoiny (1993), where a problem with ARMA errors is replaced by an independent error problem through subsequent transformations of the design space $\mathcal{X}$ requires a directional component.

The most important approaches that do carry over more easily are put into the spatial perspective below. Müller and Pázman (1999) have applied the directional derivative algorithm from this section to examples from all these approaches (and the above example) and they report very satisfactory results. For a more detailed recent survey of some of those alternative methods see Fedorov (1996). Applications for correlated error design theory and the presented according methodology cannot only be found in environmetric fields, viable input to the theory came specifically from the design of computer simulation experiments (see Sacks et al. (1989) for various setups).

Uniformly Optimum Designs

A simple direct analytic approach is possible when the regression function is generated linearly by the covariance function, i.e.

$$\dot{\eta}(x) = \sum_{x' \in A^*} c(x, x') f(x'), \quad \forall x \in \mathcal{X} \tag{5.18}$$

where A^* is some finite set and $f(x) \in \mathbb{R}^m$ are given vectors. One can regard A^* as the support of a replication-free design, in accordance with the definition in Section 5.1. These cases are considered extensively in Näther (1985b). For example, in the simple two-dimensional linear regression (5.14) with the covariance function (5.15) condition (5.18) holds for $A_9^* = \{(-1,-1), (-1,0), (-1,1), (0,-1), (0,0), (0,1), (1,-1), (1,0), (1,1)\}$ presented in Figure 5.2 and all $x \in \mathcal{X} = [-1,1]^2$, as can be further derived from

$$\begin{pmatrix} 1 \\ x_1 \\ x_2 \end{pmatrix} = c(x, x_1^*) \begin{pmatrix} 1 \\ -1 \\ -1 \end{pmatrix} + c(x, x_2^*) \begin{pmatrix} 1 \\ -1 \\ 0 \end{pmatrix} + c(x, x_3^*) \begin{pmatrix} 1 \\ -1 \\ 1 \end{pmatrix}$$

$$+ c(x, x_4^*) \begin{pmatrix} 1 \\ 0 \\ -1 \end{pmatrix} + c(x, x_5^*) \begin{pmatrix} 1 \\ 0 \\ 0 \end{pmatrix} + c(x, x_6^*) \begin{pmatrix} 1 \\ 0 \\ 1 \end{pmatrix}$$

$$+ c(x, x_7^*) \begin{pmatrix} 1 \\ 1 \\ -1 \end{pmatrix} + c(x, x_8^*) \begin{pmatrix} 1 \\ 1 \\ 0 \end{pmatrix} + c(x, x_9^*) \begin{pmatrix} 1 \\ 1 \\ 1 \end{pmatrix}.$$

Such a design A^* is uniformly optimum among all replication-free designs, i.e. it gives the full information about β, which is contained in the random process $y(x)$ (see Sacks and Ylvisaker (1966) or Appendix

A.5.5 for a detailed proof). In Pázman and Müller (1999) an algorithm based upon approximate information matrix is presented that seems particularly useful for this setting. If condition (5.18) does not hold exactly Näther (1985a) suggests to find an approximating system of functions $\{c(x, x')\}_{x' \in \mathcal{X}}$, but there is no discussion of how good a design corresponding to a specific approximation can actually be.

D-Optimum Invariant Designs

When the D-criterion is used in a linear(ized) regression, Bischoff (1993) gives conditions on the covariance matrices C and C' under which

$$|X^T C'^{-1} X| \geq |X'^T C'^{-1} X'| \text{ implies } |X^T C^{-1} X^T| \geq |X'^T C^{-1} X'^T|$$

for all possible X, X'. Setting $C' = I$ makes it possible to investigate for which correlation structures (resulting in covariance matrices C) the classical (independent observation) D-optimum designs (derived by techniques from section 3) still hold. Bischoff (1993) states that the uncorrelated exact D-optimum design corresponding to support A_D^* is D-optimum invariant (i.e. under a specific correlation scheme) if besides the inclusion condition (see Appendix A.5)

$$|C(A_D^*)| \leq |C(A)| \quad \forall A \text{ such that } \xi_{n_A} \in \Xi_{n_A} \tag{5.19}$$

holds. Note that for specific situations condition (5.19) cannot be systematically verified, except for cases where the covariance function does not depend upon the design, like

$$C(A) = I_{n_A} + \kappa \mathbf{1} \mathbf{1}^T, \qquad \kappa \in \left(-\frac{1}{n_A}, \infty\right),$$

for which any classical D-optimum regression (with intercept) design is indeed D-optimum for any such $C(\cdot)$.

Asymptotically Optimum Designs

The probably most elegant results in the domain of optimum designs when the observations are correlated have been obtained in a series of papers by Sacks and Ylvisaker (1966, 1968, 1970). Since direct calculation of optimum designs is as discussed only possible in restricted cases, they propose asymptotic Φ-optimality for sequences of designs $\{\xi_{n_A}\}$, with $n_A = 1, \ldots, \infty$, such that

$$\lim_{n_A \to \infty} \frac{\Phi[M(A)] - \Phi[M(\mathcal{X})]}{\max_{\xi_{n_A} \in \Xi_{n_A}} \Phi[M(A)] - \Phi[M(\mathcal{X})]} = 1$$

as their new design criterion. Here $\Phi[M(\mathcal{X})]$ denotes the criterion value corresponding to the total available information, i.e. when the field is observed over the whole $\mathcal{X}$. This means that they compare sequences of replication-free n-point designs according to the rate of increase in the information obtained from the corresponding experiments when the number of supporting points tends to infinity, and they subsequently construct sequences of designs which are optimal in this sense.

Their results hold basically for univariate regression on $\mathcal{X} = [0, 1]$ and under various conditions on $c(\cdot, \cdot)$, most importantly that

$$\dot{\eta}_j(x) = \int_0^1 c(x, x')\varphi_j(x')dx', \qquad j = 1, \ldots, m,$$

with continuous functions $\varphi_j(\cdot)$ (which Boltze and Näther (1982) claim to be natural for best linear unbiased estimators) and a continuous, nonzero function of jumps along the diagonal of $C(A)$, namely

$$\zeta(x) = \lim_{x' \to x^-} \frac{\partial c(x, x')}{\partial x'} - \lim_{x' \to x^+} \frac{\partial c(x, x')}{\partial x'} \geq 0.$$

Extensions of their work for less restrictive assumptions can be found in Wahba (1971), Abt (1992) and Müller-Gronbach (1996). For some examples of covariance functions fulfilling these properties, see Bellmann (1977).

In this setting (and the one-dimensional case $j = 1$) Sacks and Ylvisaker (1968) find that the asymptotically optimal sequences $\{A\}^* = \{x_1^*, \ldots, x_{n_A}^*\}$ are determined by

$$\kappa \int_0^{x_i^*} \zeta^{\frac{1}{3}}(x)[\varphi^T(x) D_\Phi \varphi(x)]^{\frac{1}{3}} dx = \frac{i}{n_A}, \qquad (5.20)$$

where κ is a norming constant and D_Φ is a fixed nonnegative square matrix related to the choice of the criterion $\Phi[\cdot]$ (in the case of D-optimality it is simply $D_\Phi = M^{-1}(\mathcal{X})$). Note the similar structure of the generating function in (5.20) and the sensitivity functions defined in Section 3.2. One of the classical examples in this context is

Brownian motion with $\dot{\eta}(x) = x^2$ and $c(x, x') = \min(x, x')$, where the sequence of equally spaced designs $x_i^* = \frac{i}{n}$, $i = 1, \ldots, n$, for $n \to \infty$ defines the asymptotically (as well as the exact) optimum n-point designs. A similar analysis for a biased estimator can be found in Su and Cambanis (1994).

Unfortunately results along this approach do not only have the practical limitation that the finite sample properties of the designs are unclear, but also that they do not readily generalize to higher (spatial) dimension. For a discussion of the difficulties see Ylvisaker (1975), who eventually proposes product designs based upon (5.20). An investigation of multidimensional equally spaced designs for a less efficient estimator under different asymptotics can be found in Bickel and Herzberg (1979).

Expansion of the Covariance Kernel

A more general approach is based on the fact that the error process in the random field (2.1) can be represented by the infinite expansion $\varepsilon(x) = \sum_{k=m+1}^{\infty} \beta_k \varphi_k(x)$, where the β_k are specific random values with $E[\beta_k] = 0$ and $E[\beta_k \beta_{k'}] = \lambda_k \delta_{k,k'} = \Lambda_{k,k'}$, and the $\varphi_k(x)$ and λ_k are the eigenfunctions and eigenvalues, respectively, of the covariance kernel $c(x, x')$ (cf. e.g. Fedorov and Hackl (1994)). Introducing auxiliary independent errors $\tilde{\varepsilon}(x)$ (but now other than in Section 5.3 not depending upon ξ) allows the basic model (2.1) to be approximated by

$$y(x) = \eta(x, \beta) + \sum_{k=m+1}^{p} \beta_k \varphi_k(x) + \tilde{\varepsilon}(x), \qquad (5.21)$$

with $E[\tilde{\varepsilon}(x)\tilde{\varepsilon}(x')] = \sigma^2 \delta_{x,x'}$ and $\varepsilon(x) - \tilde{\varepsilon}(x)$ is now assumed to have covariance $c(x, x')$. Presentation (5.21) now admits replications (as long as $\sigma^2 > 0$) and thus can be regarded as a special form of a random coefficient model (3.13) with

$$D = \begin{pmatrix} 0_{m \times m} & 0_{m \times p} \\ 0_{p \times m} & \Lambda \end{pmatrix}.$$

In this framework Fedorov and Flanagan (1997) succeed to derive the sensitivity function as

$$\phi_p(x, \xi) = \phi_\beta(x, \xi) - {}_p\varphi^T(x) \left[\frac{\sigma^2}{n} \Lambda^{-1} + \sum_{x \in \mathcal{X}} {}_p\varphi(x){}_p\varphi^T(x) \right]^{-1} {}_p\varphi(x),$$

cf. (3.14), where $_p\varphi(x) = (\varphi_1(x), \ldots \varphi_p(x))$, and a corresponding equivalence theorem. It is obvious that an optimum design derived from $\phi_p(x, \xi_*)$ will critically depend upon the order of approximation that was chosen and it is not a priori clear how large to select p. Unfortunately the most interesting limit case $p \to \infty$ is not covered by the theory and the behavior of the respective designs remains unexplored.

A Heuristic Algorithm

It is a straightforward idea to adopt algorithms from Section 3.3, developed for the independent error case, by replacing expressions in the definition of the sensitivity function by their analogous versions in the correlated setup. This suggestion goes back to Brimkulov et al. (1980), who give algorithms for some of the most widely used design criteria. E.g. for the D-criterion, where the classical sensitivity function is just the ratio of the prediction variance with the estimated parameter to the prediction variance with the true parameter (cf. Fedorov (1996)):

$$\phi_D(x, \xi_n) = \frac{E[(y(x) - \hat{y}(x))^2]}{\sigma^2(x)}$$

can be replaced by

$$\tilde{\phi}_D(x, \xi_n) = \frac{\tilde{\sigma}^2(x, \xi_n) + \tilde{\eta}^T(x, \xi_n) M^{-1}(\xi_n)\tilde{\eta}(x, \xi_n)}{\tilde{\sigma}^2(x, \xi_n)}, \qquad (5.22)$$

where

$$\begin{aligned}
\tilde{\sigma}^2(x, \xi_n) &= c(x,x) - c^T(x, \xi_n) C^{-1}(\xi_n) c(x, \xi_n), \\
\tilde{\eta}(x, \xi_n) &= \dot{\eta}(x) - \dot{\eta}(\xi_n) C^{-1}(\xi_n) c(x, \xi_n),
\end{aligned}$$

and $[c(x, \xi_n)]_i = c(x, x_i)$, $i = 1, \ldots, n$.

Originally, (5.22) was derived from the classical one-point updating formula of the determinant of a symmetric matrix. However, the above given statistical interpretation makes clear why an improvement of the design by a correction algorithm based upon $\tilde{\phi}_D(\cdot, \cdot)$ can be expected. Unfortunately there exists no proof of convergence of this heuristic algorithm (as in the uncorrelated case) and although refinements of the procedure have been suggested (see e.g. Näther (1985b)) none of them apparently guarantees perfect performance. Indeed there are many instances where in simulation experiments severe failures of finding the optimum have been reported (see Glatzer and Müller (1999)).

Nevertheless, probably due to their simplicity and their lack of restrictive assumptions on $\eta(\cdot,\cdot)$ or $c(\cdot,\cdot)$ algorithms based on this idea are the most popular in applications of correlated design problems (see e.g. Rabinowitz and Steinberg (1990)). The approach is also similar in spirit to the MEV (Minimum Estimation Variance) algorithm employed by Arbia and Lafratta (1997), although there the primary interest lies in prediction rather than parameter estimation.

5.4 Designs for Spatial Prediction

When the goal is to predict the random field at a given number of points of interest $\{_1x,\ldots,_qx\}$, it is trivial that the optimum design without restrictions on the number of support points is realized if observations are taken exactly at the sites of interest. We then have the observation we require directly and no error-stricken prediction is necessary. However, if we face the constraint that the number of observation sites must be less than the number of points of interest, i.e. $n < q$, we will have to optimize a compromise criterion, such as

$$\sum_{j=1}^{q} w(_jx)\mathrm{Var}[\hat{y}(_jx)], \qquad (5.23)$$

the weighted average variance of prediction (cf. e.g. Fedorov (1996) and (4.3)).

The universal kriging predictor is given by (2.10) and it is clear from that presentation that the quality of a design with respect to a criterion (5.23) depends highly upon the quality of both the estimated trend and the fitted variogram. Design approaches, where objectives corresponding to those two components are combined are presented in Chapter 6.

An exhaustive search over all potential designs or any other numerical minimization of the universal kriging variances (2.9) seems unfeasible in general, although some attempts for restricted settings have been made (e.g. McBratney *et al.* (1981), Cressie *et al.* (1990), recently Pesti *et al.* (1994), and Benedetti and Palma (1995)). The relative performance of the kriging estimator on various regular grids was investigated by Yfantis *et al.* (1987). The asymptotic properties for some spatial designs have been derived by Micchelli and Wahba (1981).

For more specific recommendations for design for prediction let us consider some special cases. In the uncorrelated setup the universal kriging variance (2.9) reduces to the well-known expression

$$\text{Var}[\hat{y}(_jx)] = \sigma^2(_jx)[1 + {_jx}^T(X^TX)^{-1}{_jx}],$$

or more generally

$$\text{Var}[\hat{y}(_jx)] = \sigma^2(_jx) + d[_jx, M(\xi)].$$

Therefore in this situation (5.23) can be considered a task within the framework of o.d.e. (cf. Section 4.3 for a similar case) and standard techniques from Chapter 3 may be used for the optimization.

The situation under correlated observations is more complicated. After proper detrending $\eta(x,\beta) \equiv 0$ the prediction variance in the residual field reduces to

$$\text{Var}[\hat{y}(_jx)] = \sigma^2(_jx) + c^T(_jx)C^{-1}c(_jx). \tag{5.24}$$

The possibility of detrending seems only a slightly stronger assumption than the availability of a consistent estimator $\hat{\beta}$, that is necessary for (2.7) being the best unbiased predictor in the first place. Näther (1985b) has suggested to use a design algorithm for the minimization of the prediction variance based on (5.22). Fedorov (1996) attempts to approximate the influence of the covariances on the trend by a linear correction term, which allows him to proceed in the framework of random coefficient models. He points out that (5.23) can be viewed as a special case of A-optimality and thus he can use standard design tools for its optimization. This interpretation would in principle also make the concept of approximate information matrices useful for this problem. The technicalities, however, seem too burdensome and a similarly elegant solution as in the parameter estimation case (Section 5.2) is not available. The idea of adding a regularizing (though not design dependent) noise to the random field was also used by Batsell et al. (1998), who give a promising equivalence theorem for the prediction design problem. Unfortunately their theory does not cover the interesting limit case, when the variance of this error tends to zero, and thus the performance of their approach in this case is unclear.

5.5 Example

A Replication-Free Design

The existing Upper-Austrian monitoring network consists of 17 measuring station (cf. Figure 1.1). The question, whether these stations are efficiently placed, respectively how and if relocation of these stations might improve the network is a replication-free design problem. Given a limited number of stations (here 17) and the obvious restriction that we cannot place two stations at the same spot, we are looking for an allocation that ensures 'optimal' estimation of the model parameters.

For this problem let us again assume a bivariate Gaussian p.d.f. model as in Section 3.5 with a prior estimate of the parameter vector $\beta_0 = (0.604, 0.382, 1.873, 0.854, 0.508, 0.547, -0.400)$. Figure 3.6 shows a classical locally D-optimum design, which consists of 11 support points with differing weights. It is not clear how from such a design it should be possible to derive a configuration of 17 (equally weighted) measurement stations. However, both of the two algorithms presented in Section 5.2 allow us to calculate a locally D-optimum replication-free design, which can be readily interpreted as a monitoring network. The result is presented in Figure 5.3 and it can be seen that 14 of the support points are located at the boundary of the region (within 6 principle regions) and three of them at the center. Such a design (that approximately puts equal weights at $m = 6$ vertices and central points) is in accordance with standard o.d.e. theory and the results from Section 3.5.

A Design for a Correlated Field

Let us now suppose that the errors are correlated according to the parametric (spherical) variogram estimated in Section 2.5. In the following we will assume that the covariance function (based on this estimate) is known, i.e.

$$c(x, x') = 0.487 - \gamma_S(h, (0.164, 0.323, 24.25)^T),$$

which has a range of 0.292 in the rescaled mapping. After 400 iterations (and setting $\rho = 100$) the algorithm from Section 5.3 yields a 17 point D-optimum design that is given in Figure 5.4.

This design again locates the majority of the support points along the boundary of the region, but much more evenly spaced. Also the four

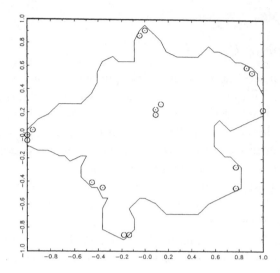

FIGURE 5.3. A locally D-optimum replication-free design.

points within the region are more spread out and altogether the design resembles the space-filling designs, which were presented in Chapter 4. In fact there are no point pairs that are particularly close to each other and neighboring points have in most cases a distance that is close to the range, which has the effect that observations from such a network are approximately independent. This is a feature that might be desirable for estimating the first order properties of a random field but which has a negative effect when the aim is the investigation of second order properties. This is discussed in Chapter 6.

Replications in Time

It has been pointed out before that air-pollution monitoring data usually (also this particular data set) carries a temporal component. Repetitive measurements at $t = 1, \ldots, T$ can be interpreted as observations from a sequence of random fields $y_t(\cdot)$. In this (different) sense the tool of replication (which was so important for the development of the theory in Chapter 3) is again made available.

In principle, there is no reason to treat the variable t differently from the spatial coordinates x and not to use an extended model (2.25). However, this would require the optimization (5.13) in the algorithm to be taken over the set $\{\bar{\mathcal{X}} \otimes [1, \ldots, T]\}$ and make the inversion of a $Tn_{\bar{\mathcal{X}}} \times Tn_{\bar{\mathcal{X}}}$ matrix necessary, which is computationally prohibitive,

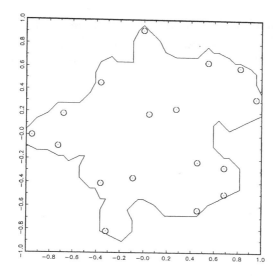

FIGURE 5.4. A locally D-optimum design under a given correlation structure.

even in the simplest cases. Also the formulation and interpretation of such an extended model (with a spatio-temporal covariance function) might pose disagreeable difficulties.

Thus it may be reasonable to assume that the y_t's are independent realizations of the random field, i.e. the covariance function is given by (2.26). The total information matrix can then be written as a double sum (see Appendix A.5.6)

$$M(A) = \frac{1}{N_A} \sum_{t=1}^{T} \sum_{x,x' \in A_t} \dot{\eta}_t(x)[C^-(A_t)]_{x,x'} \dot{\eta}_t^T(x'),$$

where $A_t = \{x_{t,i} \neq x_{t,i'}; i \neq i' = 1, \ldots, n_{A_t} \equiv N_{A_t}\}$ denotes the spatial design at time t. Consequently, we can still use the design algorithm from Section 5.3 after replacing $g(\cdot,\cdot)$ by a respective $\sum_{t=1}^{T} g_t(\cdot,\cdot)$ in (5.16) and (5.17).

For certain design criteria, namely such that $\Phi[M+M'] = \Phi[M] + \Phi[M']$, $\forall M, M'$, the analysis becomes even simpler. It is then possible to find the optimal spatial allocation for each time point and subsequently determine the optimal combination. As an example consider that we are allowed to take again 17 observations now distributed over 2 points in time. To avoid singularities we are thus required either to distribute the observations as 7 : 10 or 8 : 9 for the two dates. As

118   5. Designs for Spatial Trend Estimation

a design criterion consider $\Phi[M] = \text{tr} M$, which is additive. We can therefore construct the designs separately and then base the decision on which combination to take on the sum of the criterion values.

n_A	$\text{tr} M(A)$
7	37.46
8	43.52
9	48.45
10	53.94

TABLE 5.1. Separately evaluated design criteria.

Table 5.1 gives the criterion values for the separately evaluated designs (the best out of 10 runs for each n_A). It is easy to see that the combination 8 : 9 is slightly preferable (91.97 > 91.40) and it indeed gives a design with replications as can be seen from Figure 5.5.

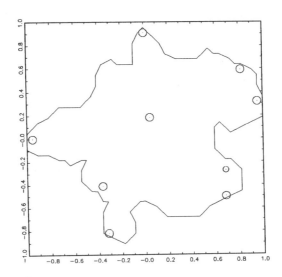

FIGURE 5.5. A Φ-optimum design under a given correlation structure with replications in time (the smaller circle indicates an unreplicated point).

5.6 Exercises

1. By the two methods given (and for both the nonlinear model selected and the second order polynomial) calculate 36 point replication-free designs. Are the results in accordance?

2. Calculate 36 point replication-free designs for

    - localizing the peak,
    - estimating the mean parameter in a random coefficient setting,
    - discriminating the models.

3. Assuming the correlation structure is according to your preestimated variogram. Construct an optimum design by the approximate information matrix algorithm and a Brimkulov et al. (1980) type procedure, and compare their efficiencies.

4. For all points in $\mathcal{X}$ find a design that minimizes the average prediction variance (5.23) by assuming first uncorrelated errors and later relaxing this assumption.

5. Let the 36 observations be distributed on 2 (later more) points in time. Find a design that maximizes a reasonable criterion. Are there any replications?

Connections to Other Chapters

This chapter brought a new point of view towards the optimum design problems for replication-free experiments and for correlated observations, since there seems to be no other work that exploits the concept of design measures in this context. By introducing approximate information matrices and an additional white noise dependent upon the design, we were able to make the design problem smooth in such a way that standard techniques of differentiation could be used. This allowed us to elaborate a gradient algorithm that is of potential use in applications, especially in complex settings, where enumerative methods are unfeasible.

A critical assumption for all the methods in this chapter was that the covariance structure is known in advance. There remains the open problem of how to construct designs that can be efficiently employed for estimating this covariance structure (or in a more specific context the variogram). Eventually a combination of such designs and the designs presented in this chapter are needed. This will be the main topic of the final Chapter 6.

References

Abt, M. (1992). Some exact optimal designs for linear covariance functions in one dimension. *Communications in Statistics, Theory & Methods*, 21:2059–2069.

Arbia, G. and Lafratta, G. (1997). Evaluating and updating the sample design in repeated environmental surveys: monitoring air quality in Padua. *Journal of Agricultural, Biological, and Environmental Statistics*, 2(4):451–466.

Batsell, S., Fedorov, V. and Flanagan, D. (1998). Multivariate prediction: Selection of the most informative components to measure. *MODA5*, Atkinson, A.C., Pronzato, L., and Wynn, H.P., editors. Physica Verlag, Heidelberg.

Bellmann, A. (1977). Zur Konstruktion asymptotisch optimaler Folgen von Versuchsplänen bei korrelierten Beobachtungen (in German). *Freiburger Forschungshefte*, D 106:79–91.

Benedetti, R. and Palma, D. (1995). Optimal sampling designs for dependent spatial units. *Environmetrics*, 6:101–114.

Bickel, P.J. and Herzberg, A.M. (1979). Robustness of designs against autocorrelation in time I: Asymptotic theory, optimality for location and linear regression. *The Annals of Statistics*, 7(1):77–95.

Bischoff, W. (1993). On D-optimal designs for linear models under correlated observations with an application to a linear model with multiple response. *Journal of Statistical Planning and Inference*, 37:69–80.

Boltze, L. and Näther, W. (1982). On effective observation methods in regression models. *Mathematische Operationsforschung und Statistik, Series Statistics*, 13(4):507–519.

Brimkulov, U.N., Krug, G.K. and Savanov, V.L. (1980). Numerical construction of exact experimental designs when the measurements are correlated (in Russian). *Zavodskaya Laboratoria (Industrial Laboratory)*, 36:435–442.

Conan-Doyle, A. (1904). The adventure of the golden pince-nez. In *The Return of Sherlock Holmes*. Strand Magazine, London.

Cook, D. and Fedorov, V. (1995). Constrained optimization of experimental design (with discussion). *Statistics*, 26:129–178.

Cressie, N., Gotway, C.A. and Grononda, M.O. (1990). Spatial prediction from networks. *Chemometrics and Intelligent Laboratory Systems*, 7:251–271.

Fedorov, V.V. and Flanagan, D. (1997). Optimal monitoring network design based on Mercer's expansion of covariance kernel. *Journal of Combinatorics, Information & System Sciences*, 23:237–250.

Fedorov, V.V. and Hackl, P. (1994). Optimal experimental design: Spatial sampling. *Calcutta Statistical Association Bulletin*, 44:57–81.

Fedorov, V.V. (1989). Optimal design with bounded density: Optimization algorithms of the exchange type. *Journal of Statistical Planning and Inference*, 22:1–13.

Fedorov, V.V. (1996). Design of spatial experiments: model fitting and prediction. In Gosh, S. and Rao, C.R., editors, *Handbook of Statistics*, volume 13. Elsevier, Amsterdam, 515–553.

Glatzer, E. and Müller, W.G. (1999). A comparison of optimum design algorithms for regressions with correlated observations. In Pázman, A., editor, *Probastat 98*, Tatra Mountains Mathematical Publishers, 17:149–156.

Kiefer, J. (1959). Optimal experimental designs (with discussion). *Journal of the Royal Statistical Society, Series B*, 272–319.

McBratney, A.B., Webster, R. and Burgess, T.M. (1981). The design of optimal sampling schemes for local estimation and mapping of regionalized variables — I. *Computers and Geosciences*, 7:331–334.

Micchelli, C.A. and Wahba, G. (1981). Design problems for optimal surface interpolation. In *Approximation Theory and Applications*, Academic Press, 329–349.

Müller, W.G. (1995). An example of optimal design for correlated observations (in German). *Österreichische Zeitschrift für Statistik*, 24(1):9–15.

Müller, W.G. and Pázman, A. (1998). Design measures and approximate information matrices for experiments without replications. *Journal of Statistical Planning and Inference*, 71:349–362.

Müller, W.G. and Pázman, A. (1999). An algorithm for the computation of optimum designs under a given covariance structure. *Computational Statistics*, 14(2):197–211.

Müller-Gronbach, T. (1996). Optimal designs for approximating the path of a stochastic process. *Journal of Statistical Planning and Inference*, 49:371–385.

Näther, W. (1985a). *Effective Observation of Random Fields*. Teubner Texte zur Mathematik — Band 72. Teubner Verlag, Leipzig.

Näther, W. (1985b). Exact designs for regression models with correlated errors. *Statistics*, 16(4):479–484.

Pázman, A. (1986). *Foundations of Optimum Experimental Design*. Mathematics and Its Applications. D.Reidel, Dordrecht.

Pázman, A. and Müller, W.G. (1998). A new interpretation of design measures. In *Model-Oriented Data Analysis 5*, Atkinson, A.C., Pronzato, L., and Wynn, H.P., editors. Physica Verlag, Heidelberg.

Pázman, A. and Müller, W.G. (1999). Properties of design measures for optimal observations from correlated processes, Report No. 411 of the Institute of Mathematics of the University Augsburg.

Pázman, A. and Müller, W.G. (2000). Optimal Design of Experiments Subject to Correlated Errors, *Statistics & Probability Letters*, forthcoming.

Pesti, G., Kelly, W.E. and Bogardi, I. (1994). Observation network design for selecting locations for water supply wells. *Environmetrics*, 5:91–110.

Pukelsheim, F. (1993). *Optimal Design of Experiments*. John Wiley & Sons, Inc., New York.

Rabinowitz, N. and Steinberg, D.M. (1990). Optimal configuration of a seismographic network: A statistical approach. *Bulletin of the Seismological Society of America*, 80:187–196.

Rasch, D.A.M.K., Hendrix, E.M.T. and Boer, E.P.J. (1997). Replicationfree optimal designs in regression analysis. *Computational Statistics*, 12:19–32.

Sacks, J., Welch, W.J., Mitchell, T.J. and Wynn, H.P. (1989). Design and analysis of computer experiments. *Statistical Science*, 4(4):409–435.

Sacks, J. and Ylvisaker, D. (1966). Design for regression problems with correlated errors. *Annals of Mathematical Statistics*, 37:66–89.

Sacks, J. and Ylvisaker, D. (1968). Design for regression problems with correlated errors; many parameters. *Annals of Mathematical Statistics*, 39:46–69.

Sacks, J. and Ylvisaker, D. (1970). Design for regression problems with correlated errors III. *Annals of Mathematical Statistics*, 41:2057–2074.

Saunders, I.W. and Eccleston, J.A. (1992). Experimental design for continuous processes. *Australian Journal of Statistics*, 34(1):77–89.

Schilling, M.F. (1992). Spatial design when the observations are correlated. *Communications in Statistics — Simulation*, 21(1):243–267.

Spokoiny, V.G. (1993). Optimal design of regression experiments with ARMA-errors. *Mathematical Methods of Statistics*, 2(1):45–61.

Su, Y. and Cambanis, S. (1994). Sampling designs for regression coefficient estimation with correlated errors. *Annals of the Institute of Statistical Mathematics*, 46:707–722.

Wahba, G. (1971). On the regression design problem of Sacks and Ylvisaker. *Annals of Mathematical Statistics*, 42:1035–1053.

Wynn, H.P. (1982). Optimum submeasures with applications to finite population sampling. In *Statistical Decision Theory and Related Topics III*, Academic Press, 2:485–495.

Yfantis, E.A., Flatman, G.T. and Behar, J.V. (1987). Efficiency of kriging estimation for square, triangular, and hexagonal grids. *Mathematical Geology*, 19:183–205.

Ylvisaker, D. (1975). Design on random fields. *A Survey on Statistical Design and Linear Models*, 593–607.

Yuditsky, M.I. (1974). About optimal design in estimation of parameters of regression models with correlated errors of observations (in Russian). *Theorija Verodjatnosteij i Matematnjeckaja Statistika*, 11:169–176.

6
Multipurpose Designs Including Designs for Variogram Fitting

> *"[...] Ain't no particular sign I'm all compatible with, [...]"*
> Prince and the Revolution (1986)

At many points of this monograph (especially in Chapter 4) it became clear that a single purpose design may be quite inefficient for handling a real-life problem. Therefore we often need to incorporate more than one design criterion and a common approach is simply to construct a weighted average

$$\bar{\Phi}[\xi|\alpha] = \alpha\Phi[M(\xi)] + (1-\alpha)\Phi'[M'(\xi)], \qquad (6.1)$$

(cf. Läuter (1976)), which may depend upon different information matrices M and M'. The weighting parameter $0 \leq \alpha \leq 1$ has to be selected by the user and it is not very clear (due to the generally different scaling of $\Phi[\cdot]$ and $\Phi'[\cdot]$), which choice of α corresponds to an intended weighting. The weighted average (6.1) is used similarly to the criteria described in Section 3.2. An extension to more than two criteria is straightforward. Related ideas for combining the purposes of parameter estimation and model discrimination already appear in Fedorov (1972). In certain situations it is possible to adopt methods from Chapter 3 for their optimization (see e.g. Cook and Nachtsheim (1982), or Dette (1990)). Designs based upon (6.1) and its straightforward generalization to more than two terms have been termed compound designs; they also prove useful for the situation of multiple (e.g. pollution or seismic) sources influencing the response (see e.g. Steinberg et al. (1995)).

Another method to satisfy multiple design criteria is the one of constrained optimum designs (for a recent survey, see Cook and Fedorov (1995)). Instead of (6.1) one could for instance choose to find

$$\xi^* = \arg\max_{\xi \in \Xi} \Phi[M(\xi)] \quad \text{s.t.} \quad \Phi'[M'(\xi)] > \kappa,$$

i.e. an optimum design for criterion $\Phi[\cdot]$ which ensures sufficiently good estimation with respect to the second criterion $\Phi'[\cdot]$ (or vice versa). A disadvantage of constrained optimum designs is the unsymmetric involvement of the considered goals. The relationship between constrained and compound designs is explored in detail in Cook and Wong (1994).

The need to satisfy more than one design criterion is particularly relevant in the context of random fields. Recalling the discussion in Section 2.4 it is evident that for precise universal kriging it is important not only to efficiently estimate the spatial trend parameters β, but also the parameters θ of the variogram or covariance function. Both tasks could for instance be comprised by applying corresponding design criteria $\Phi[M(\cdot,\beta)]$ and $\Phi'[M'(\cdot,\theta)]$ and constructing a compound design from $\bar{\Phi}[\cdot]$. Design criteria that are suitable for estimation of second order characteristics will be explored in the following section. Techniques, such as (6.1) and its alternatives, for combining designs for first and second order characteristics will be suggested in the latter sections of this chapter.

Note that in the case of independent errors (and heterogeneous error variances) a 'naive', and more direct design approach to the compound problem is possible. As an example consider a simple linear regression $y(x) = \beta_1 + \beta_2 x + \varepsilon(x)$, $x \in [-1, 1]$, with a parameterized variance $\sigma^2(x) = |x|^{2\theta}$, with $\theta \geq 0$. We can then simply 'divide' the original model by the standard deviation to arrive at the nonlinear model

$$y'(x) = \beta_1 |x|^{-\theta} + \beta_2 |x|^{1-\theta} \text{sgn}(x) + \varepsilon', \qquad \forall x \neq 0,$$

with homoscedastic errors ε'. It is then possible to apply standard methods from Chapter 3 to find optimum designs for the estimation of β and θ. The theoretical properties of this 'naive' technique and its limitations are given by Maljutov (1988), a theoretically more rigorous approach based upon the full specification of the information matrix is given in Atkinson and Cook (1995).

In the example above a curious feature of the inclusion of a variance parameter is that it implies that the D-optimum design will become unsymmetric, depending upon the relative size of the intercept and the slope of the linear regression even under the assumption of homoscedasticity. This can be seen from Figure 6.1, which gives a local design when $\theta_0 = 0$. The solid line depicts the sensitivity function and the bars stand for the D-optimum design measure for the prior guesses $\beta_0^T = (1, 10)$. The optimum design becomes a more symmetric 4-point design when β_{02} is increased relatively to β_{01} and collapses to a uniform three point design supported on $\{-1, 0, 1\}$ when the size of the intercept dominates the slope. Remember, that the classical D-optimum design for linear regression is to put an equal portion of the measure at the vertices of the design region.

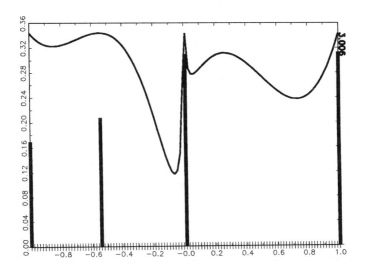

FIGURE 6.1. A D-optimum design for linear regression under heteroscedasticity.

6.1 Designs for Variogram Estimation

As was already emphasized in Chapter 2 the variogram plays a central role in the analysis of geostatistical data. A valid variogram model is selected, and the parameters of that model are estimated, before kriging (spatial prediction) is performed. These inference procedures are generally based upon examination of the empirical variogram, which consists of average squared differences of data taken at sites lagged the same distance apart in the same direction (cf. (2.21)). The ability of an investigator to estimate the parameters of a variogram model efficiently is affected significantly by the sampling design, i.e., the locations of the sites $x_1, \ldots, x_n \in \mathcal{X}$ where data y are taken. For instance, if $y(\cdot)$ is isotropic and no two sites are within a distance $\epsilon > 0$ of each other, then it will be difficult, if not impossible, to estimate precisely those parameters whose primary influence is on the behavior of the variogram for distances less than ϵ. Stein (1988) has shown that for efficient prediction, proper specification of the variogram is much more important for 'small' distances than for 'large' distances. The influence of data sparsity (and the lack of small distances) on variogram estimation is well documented by Lamorey and Jacobson (1995).

Because kriging is usually the ultimate objective of a geostatistical analysis, previous authors who have considered design aspects in this context have emphasized the utility of designs for prediction, under the assumption that both the functional form of the variogram and the values of its parameters are known; see Section 5.4. Inasmuch as the variogram must in practice be estimated, however, a total emphasis on prediction with no regard to the utility of the design for variogram estimation seems to 'put the cart before the horse'. In Müller and Zimmerman (1999) several practical approaches for constructing sampling designs for estimation of the variogram were compared by Monte-Carlo simulations. Those designs could be adopted at the early stages of a sampling program until the variogram is sufficiently well-estimated, after which one could shift to an existing approach that emphasizes prediction. Alternatively, the two objectives could be comprised in a compound design, as suggested in Section 6.4.

Pettitt and McBratney (1993) have suggested designs for variogram estimation in the absence of any knowledge about the extent of variation of the noise (very much in the spirit of Chapter 4). Such designs could be used as exploratory or starting designs for the algorithms presented in the next sections. Here we require some prior knowledge about the form and magnitude of local variation. Such knowledge was also taken into account by Russo (1984) and Warrick and Myers (1987) in their criteria for variogram estimation. Russo considered a design to be good if it minimized the dispersion of lags (i.e., displacements between the sites $x_1, \ldots, x_n$) within lag classes, while Warrick and Myers proposed a criterion that measures how well the lag distribution corresponding to a design conforms to a prespecified distribution. Both have been critized by Morris (1991) for not taking into account the correlated nature of the classical variogram estimates. The approaches presented here deal with this issue. But more fundamentally, we will investigate what that underlying prespecified lag distribution should be depending upon our prior knowledge about the variogram.

Differences to Standard Design Problems

Many aspects of the design approach proposed in this section are adaptations and/or modifications of existing approaches for trend estimation (cf. Chapter 3) to the peculiarities of the purpose of variogram fitting. A decisive difference between the two contexts (trend and variogram estimation) is the 'order' of the properties of the random process which the models characterize. In the regression context attention focuses on the expectation function, i.e., the first-order properties of the process. In the present context primary interest lies in the variogram, which characterizes the second-order properties. Design criteria are proposed that, in contrast to the design criteria oriented towards trend estimation (or kriging with a known variogram), emphasize the accurate modelling and estimation of the variogram. The design for estimating second order properties has gained considerable interest in other areas of statistics as well, e.g. computer simulation experiments (see e.g. Sacks *et al.* (1989)).

One of the key features that distinguishes this design problem from similar design problems in the standard regression setting is that the addition of a new site to an existing design of n sites produces not one but n new lags and hence many more squared differences from which the variogram is estimated. Any sensible design criterion should take account of all the new lags that are produced by a new site. Secondly, the corresponding differences are generally correlated, cf. e.g. (2.22), which inhibits the use of many standard design methods, that rest upon the assumption of uncorrelated observations.

Additionally, as can be seen from the presentation in Section 2.1, variogram models are usually nonlinear in the parameters and again the problem of the dependence of the information matrix upon the values of the unknown parameters (see the discussion in Section 3.2) arises. In particular, the information matrix M' that corresponds to $\tilde{\theta}$ from (2.24) is given by

$$M'(\xi_n, \theta) = \dot{\gamma}(\xi_n, \theta)\Sigma^{-1}(\xi_n, \theta)\dot{\gamma}^T(\xi_n, \theta), \qquad (6.2)$$

where the $p \times N^\#$ matrix

$$\dot{\gamma}(\xi_n, \theta) = \frac{\partial \gamma(h_i, \theta)}{\partial \theta_j}\Big|_\theta \quad (i = 1, \ldots, \binom{n}{2}) \equiv N^\#; \quad j = 1, \ldots, p),$$

and $\Sigma(\xi_n, \theta)_{ik} = \text{Cov}(\hat{\gamma}_i, \hat{\gamma}_k)$ is defined by a parameterized version of (2.22). Equivalently to Section 3.2 we avoid a circular problem by replacing θ by a preliminary guess θ_0 and finding a 'local' solution. All the expressions in the next section will be dependent upon θ_0 and this argument will be omitted where it leads to no ambiguity.

6.2 Augmenting Designs

Augmenting an Existing Design by One Site

Let us first assume that data from a design of n distinct sites are available. We might then face the problem of deciding where to put one additional site in order to retrieve the maximum amount of information about the parameter θ. This problem was first treated in the context of variogram estimation (under the assumption of uncorrelated observations) by Zimmerman and Homer (1991). They suggested to employ the D-criterion, i.e. to maximize of the determinant of the information matrix

$$x^* = \arg\max_{x \in \mathcal{X}_{n+1}} |M'(\xi_{n+1})|, \qquad (6.3)$$

where $S_{\xi_{n+1}} = S_{\xi_n} \cup \{x\}$ with a corresponding reweighting and $\mathcal{X}_{n+1} = \mathcal{X} \setminus S_{\xi_n}$. Straightforward optimization of (6.3) leads to an optimization problem that requires the inversion of the $\binom{n+1}{2} \times \binom{n+1}{2}$ dimensional matrix $\Sigma(\xi_{n+1}, \theta_0)$ in (6.2) for each candidate point. That the information matrix in this case is still a useful basis for a design criterion is justified by Abt and Welch (1998).

In the correlated case, it follows from well-known matrix inversion and determinant formulae (see Appendix A.6.1) that an equivalent optimization problem to (6.3) is

$$x^* = \arg\max_{x \in \mathcal{X}_{n+1}} |V(x,\xi_n)| \times \qquad (6.4)$$
$$|V^{-1}(x,\xi_n) + G^T(x,\xi_n)[\dot{\gamma}(\xi_n)\Sigma^{-1}(\xi_n)\dot{\gamma}^T(\xi_n)]^{-1}G(x,\xi_n)|,$$

where

$$G(x,\xi_n) = \dot{\gamma}(\xi_n)\Sigma(\xi_n)^{-1}\Sigma(x,\xi_n) - \dot{\gamma}(x),$$

and

$$V(x,\xi_n) = [\Sigma(x) - \Sigma(x,\xi_n)^T \Sigma(\xi_n)^{-1} \Sigma(x,\xi_n)]^{-1}$$

from the block decomposition

$$\Sigma(\xi_{n+1},\theta_0) = \begin{pmatrix} \Sigma(\xi_n) & \Sigma(x,\xi_n) \\ \Sigma(x,\xi_n)^T & \Sigma(x) \end{pmatrix}.$$

Here the $n \times n$ matrix $\Sigma(x)$ denotes the covariance matrix of the observations corresponding to the lags between the point x and the existing sites, and consequently the $\binom{n}{2} \times n$ matrix $\Sigma(x,\xi_n)$ stands for the covariance matrix of observations from these lags and all lags amongst the existing sites.

Rule (6.4) requires the inversion of the $\binom{n}{2} \times \binom{n}{2}$ matrix $\Sigma(\xi_n)$ only once, and for each candidate point just the inversion of the $n \times n$ matrix $V(x,\xi_n)$ is needed.

Augmenting Several Sites

In principle it seems straightforward to extend the design method from the previous section to choose the "optimal" locations for some additional $q > 1$ sites.

The best way to accomplish this is to take a sequential data-adaptive approach; that is, for $i = n, n+1, \ldots, n+q-1$, the optimal location for the $(i+1)$th site shall be determined using the data from the existing i sites and rule (6.4). This sequential approach permits the continual updating of the GLS estimate of θ, which is important for the reasons mentioned in Section 2.2. However, in most applications the sequential gathering of information is not possible and a one-shot design is required. Then the advantage of seizing the maximum amount of information possible at every step is lost and we are left with an algorithm (which Oehlert (1997) characterizes as 'greedy') of the form

$$x_i^* = \arg\max_{x \in \mathcal{X}_{i+1}} |V(x,\xi_i)| \times \qquad (6.5)$$
$$|V^{-1}(x,\xi_i) + G^T(x,\xi_i)[\dot{\gamma}(\xi_i)\Sigma^{-1}(\xi_i)\dot{\gamma}^T(\xi_i)]^{-1}G(x,\xi_i)|,$$

where $i = n, n+1, \ldots, n+q-1$.

It is much simpler computationally to optimize the location of a single site, one at a time. Nevertheless, the algorithm can be extended to permit the addition of more than one site (even all q sites), at a time. However, for typical applications, this is even for current computing resources, prohibitive.

The single point correction algorithm (6.5) is an adaptation of the algorithm by Brimkulov et al. (1980), for regression designs with correlated observations (see Section 5.4). For a similar suggestion in the current context see also Bogaert and Russo (1999). It is simple to extend (6.5) to a Fedorov exchange-type algorithm like in Section 3.3, as follows: at every step s add the point x_s^* to the current design $\xi_{(s)n}$ and delete the point

$$x_s^- = \arg\min_{x \in S_{\xi_{(s)n}}} |V(x, \xi_{(s,x)n})||V^{-1}(x, \xi_{(s,x)n}) +$$
$$G^T(x, \xi_{(s,x)n})[\dot{\gamma}(\xi_{(s,x)n})\Sigma^{-1}(\xi_{(s,x)n})\dot{\gamma}^T(\xi_{(s,x)n})]^{-1}G(x, \xi_{(s,x)n})|,$$

where $S_{\xi_{(s,x)n}} = \{S_{\xi_{(s)n}} \setminus x\}$.

Unfortunately, as mentioned before, due to the presence of correlations there is no guarantee (unlike to the classical uncorrelated case — diagonal Σ) that such a procedure converges and though we have empirical evidence that the improvements by it are considerable, eventually we might be trapped at a local optimum.

There have been several other attempts to solve the design problem in the correlated errors setting, for instance Sacks and Ylvisaker (1966), Fedorov and Flanagan (1997), cf. Section 5.4. All those methods seem unfeasible for the current problem, mainly due to the complex structure of the covariance function and the consequent difficulties meeting the various assumptions taken in these approaches. The method based upon approximate information matrices from Section 5.3 for instance requires nonsingular covariance matrices defined on the whole collection of candidate points (which might hardly be the case in geostatistical practice).

6.3 Alternative Methods which Ignore Correlations

A major disadvantage of the augmentation algorithms given in the previous section is that they can be computationally very demanding.

6.3 Alternative Methods which Ignore Correlations   133

Even the simplified version (6.5) might be a computational burden in some settings.

Suppose, however, that the spatial configuration of sampling locations is nearly such that sites can be grouped into pairs, with each location lying a distance of at least $h_{\min} > a$ (a denotes the variogram range) from every other location, except possibly the one with which it is paired. Then the contributions to the $\hat{\gamma}(h_{ij})$ corresponding to distinct pairs are uncorrelated with each other (though not generally homoscedastic), hence one could fit a model to the $\frac{n}{2}$ corresponding points in the (semi-)variogram cloud (ignoring the others) by weighted least squares without loosing efficiency. Thus we can now take up Cressie's (1985) suggestion to approximate Σ by a diagonal matrix, which speeds up computations considerably (as was done by Zimmerman and Homer (1991)). Note, however that the effect of ignoring large correlations (due to small lag distances) might be substantial.

Moreover, the restriction to the framework of uncorrelated observations allows us to develop alternatives to (6.5) that avoid some of its disadvantages. A two stage strategy was first suggested by Müller and Zimmerman (1995):

(a) To find the optimal configuration of distances $\xi_{\mathcal{L}}^*$ in the space spanned by all possible point pair distances (the so-called lag space $\mathcal{L}$),

(b) and to map this configuration into the original site space $\mathcal{X}$ (find a site space design).

A detailed description of such a procedure is the subject of the remaining part of this section. This approach has the advantage that (a) is in fact the solution of a classical design problem and many algorithms with guaranteed convergence are available, see Section 3.3.

Specifically, we shall subsequently pretend that the $\hat{\gamma}(h_{ij})$'s follow the model
$$\hat{\gamma}(h) = \gamma(h, \theta) + \varepsilon(h),$$
where $\gamma(h, \theta)$ is a valid variogram, specified up to the value of an unknown parameter vector θ, and the errors ε are uncorrelated and satisfy $\text{Var}[\varepsilon(h)] \propto \gamma^2(h, \theta)$. We then estimate θ by minimizing the weighted residual sum of squares by an iterative procedure from the observations from the distinct site pairs.

For a spherical variogram model (see Figure 2.2), for instance, the corresponding information matrix (cf. Müller and Zimmerman (1995)) is

$$M'(\xi_\mathcal{L}, \theta) = \frac{1}{(\theta_1 + \theta_2)} \sum_{h \in [\theta_3, \max h]} \begin{pmatrix} 1 & 1 & 0 \\ 1 & 1 & 0 \\ 0 & 0 & 0 \end{pmatrix} \xi_\mathcal{L}(h) + \quad (6.6)$$

$$\sum_{h \in [0, \theta_3]} \begin{pmatrix} 1 & \frac{h}{2\theta_3}(3 - \frac{h^2}{\theta_3^2}) & \frac{3\theta_2 h}{2\theta_3^2}(\frac{h^2}{\theta_3^2} - 1) \\ \frac{h}{2\theta_3}(3 - \frac{h^2}{\theta_3^2}) & \frac{h^2}{4\theta_3^2}(3 - \frac{h^2}{\theta_3^2})^2 & \frac{3\theta_2 h^2}{4\theta_3^3}(3 - \frac{h^2}{\theta_3^2})(\frac{h^2}{\theta_3^2} - 1) \\ \frac{3\theta_2 h}{2\theta_3^2}(\frac{h^2}{\theta_3^2} - 1) & \frac{3\theta_2 h^2}{4\theta_3^3}(3 - \frac{h^2}{\theta_3^2})(\frac{h^2}{\theta_3^2} - 1) & \frac{9\theta_2^2 h^2}{4\theta_3^4}(\frac{h^2}{\theta_3^2} - 1)^2 \end{pmatrix}$$

$$\gamma_S^{-2}(h, \theta)\xi_\mathcal{L}(h).$$

It is unfortunate that in this model (as in all nonlinear models) $M'(\cdot)$ depends upon the true values of the parameters, with the consequence that all scalar design criteria based on $M'(\cdot)$ are affected. However, the determination of solutions for various guesses allows one to assess whether the design is robust with respect to the values of the parameters θ. It is straightforward to apply the concepts from Chapter 3 on equation (6.6). A generic optimum design for the spherical variogram (in the lag space $\mathcal{L}$) is presented in Figure 6.2. It turns out that a design consisting of an equal amount of distances for very small lags (close to $h = 0$), lags larger than the range $a = \theta_3$ and intermediate lags are optimal and thus the dependence upon the local guess θ_0 is severe only for this last group of lags and has a relatively small overall effect.

Finding the solution of (b) was also the purpose of Warrick and Myers (1987). They suggest to find an optimum site space design by minimizing (a categorized version of)

$$\sum_{i=1}^{N^\#} w_i (h_i^* - h_i)^2,$$

where h_i denotes the ordered sequence of lags resulting from a design ξ, the h_i^* the corresponding sequence for an optimum design in the lag space $\xi_\mathcal{L}^*$, and w_i are prespecified weights reflecting the importance of certain lags. This rule is basically what is known as *least squares multidimensional scaling* (see e.g. Cox and Cox (1994)). In fact we may employ a number of techniques developed in that field for the solution

6.3 Alternative Methods which Ignore Correlations 135

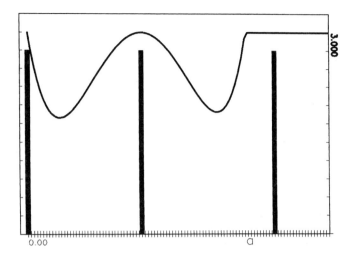

FIGURE 6.2. A generic D-optimum design for a spherical variogram in the lag space (bars stand for design weights p_i and the curve represents the sensitivity function).

of (b). Warrick and Myers (1987) suggest the use of a uniform target distribution $\xi_{\mathcal{L}}^*(\cdot)$, which is not a particular good choice as we saw in the above example. This example rather suggests a distribution which is highly concentrated at three distinct points.

There is still one open question related to (b). Since (a) is based on the assumption of uncorrelated observations it would be correct to separate the $\frac{n}{2}$ optimal lags, such that the remaining $\frac{n(n-2)}{2}$ lags are distant enough so not to produce correlations, i.e. minimizing

$$\sum_{h_i \in L_{\frac{n}{2}}} w_i(h_i^* - h_i)^2, \quad \text{s.t. } h_i \notin L_{\frac{n}{2}} > a,$$

where $L_{\frac{n}{2}}$ denotes the set of points corresponding to the $\frac{n}{2}$ optimal lags. This can be accomplished using techniques from Lee (1984) or the algorithm given below. It would guarantee an optimal (or nearly optimal) allocation. However, it is not clear if this advantage outweighs the deliberate sacrifice of a very large portion of the observations. In Müller and Zimmerman (1999) this approach is compared to an alternative, which utilizes the information from all observations, by Monte-Carlo simulations; no significant differences were found.

A (Distance) Algorithm

The proposed algorithm can be best described as a method of putting a collection of pins of various lengths on a given surface, say a table, in such a way that the distances between them are as large as possible. In other words, for a given number n of possible original sites, we can define $\frac{n}{2}$ line segments with lengths and frequencies approximately corresponding to the optimum design $\xi_\mathcal{L}^*(h)$ derived by standard methods from the previous section.

(0) Initialize the procedure by randomly distributing the segments within the region $\mathcal{X}$, thereby assigning endpoints $x_i = (x_{i1}, x_{i2})$ and $x_i' = (x_{i1}', x_{i2}')$ to each segment.

(1) Construct the matrix $\mathcal{H} = \mathcal{H}_{kl}, k = 1, \ldots, n, l = 1, \ldots, n-2$ with the minimum distance between all endpoints, i.e. with the entries being the smallest value of $\sqrt{(x_{i1} - x_{j1})^2 + (x_{i2} - x_{j2})^2}$, $\sqrt{(x_{i1} - x_{j1}')^2 + (x_{i2} - x_{j2}')^2}$, $\sqrt{(x_{i1}' - x_{j1})^2 + (x_{i2}' - x_{j2})^2}$, and $\sqrt{(x_{i1}' - x_{j1})^2 + (x_{i2}' - x_{j2})^2}$ (for $i \neq j$). Find the smallest entry $h_{\min} = \min \mathcal{H}_{kl}$ and identify the two points x_k^* and x_l^* corresponding to it.

(2) Select randomly one of the two points x^* and use its counterpart x'^* as an anchor. Randomly select a point $x^{\#}$ within $\mathcal{X}$ that lies approximately the same distance from x'^* as x^* does (lying on a circle centered at x'^*).

(3) Check if the distance to the closest of the other points is larger than $h_{\min}$. If yes, exchange x^* by $x^{\#}$; if no, repeat (2) until all points on the circle have been checked.

(4) Repeat (1) until either $h_{\min} > a$ or $h_{\min}$ can not be further increased.

Note that the algorithm provides a measure of how well the independence criterion is met, namely $\hat{\kappa} = \frac{\gamma(h_{\min})}{\gamma(a)}$, which is evidently $\kappa = 1$ in case of complete independence and $\kappa < 1$ else.

There are several factors that might affect the applicability of the given approach. The region might be too small to accommodate the independent point pairs and their respective distances, i.e. the 'independence' measure $\hat{\kappa}$ depends not only upon a, but also the size (and shape) of the region and the number of sites n.

Since classical optimum design theory assumes independent observations, the location of point pairs in the observation space has to guarantee independence. This can only be achieved if the variogram has a finite range and the original space is large enough with respect to this range and the number of original observations. The algorithm presented here separates point pairs as far as possible, thus giving the closest approximation to (or, ideally meeting the condition of) independence. Note that instead of using the information of $\frac{n(n-1)}{2}$ possible observations only $\frac{n}{2}$ are used here.

The usefulness of the algorithm can only be assessed in two ways: via simulation, or by comparing it to a technique that directly employs ideas for optimum designs for correlated observations. First results in this direction can be found in Müller and Zimmerman (1999). They conclude that algorithms based on ignoring correlations are much quicker but marginally less efficient with respect to a design criterion than the augmentation procedures from the previous section.

6.4 Combining Different Purpose Designs

For an overall efficient design it is necessary to combine optimum designs for trend estimation from Chapter 5 and for variogram estimation from the previous sections. Simultaneous optimization (in the site and the lag space) according to (6.1) is an unfeasible task, even with today's computer technology. A straightforward alternative is to firstly construct an optimum design for trend estimation with $n_0 < n$ sites by applying the algorithm from Section 5.3 and then to use the augmentation rule for the $n - n_0$ remaining sites to cover variogram fitting.

What seems of decisive importance is a good choice of n_0, the number of sites to be allocated to the trend estimation design. Warrick and Myers (1987) suggested taking $n_0 = \frac{n}{2}$, but this seems excessive and, because of the relative importance of small lags on variogram estimation, can cause the remaining $n - n_0$ optimally chosen sites to be concentrated very closely to one another.

The choice $n_0 \leq 0.3n$ seems to be better (at least for large n), a rough justification of which is the following (see also Müller and Zimmerman(1997)): Take n to be fixed, and suppose a regular grid of n_0 sites (where n_0 is to be determined) are to be used in a starting design. Let us classify a lag as "large" or "small" according to whether it is greater than or less than ϵ, where ϵ is a positive constant smaller than the grid spacing of a regular grid of n sites. Then whatever the value of n_0, all $\binom{n_0}{2}$ lags in the starting design are large and there is a dire need for small lags. The most efficient way to produce small lags is to take all of the $n - n_0$ remaining sites to be within ϵ of each other and within ϵ of an arbitrary site in the starting design. This results in $\binom{n-n_0+1}{2}$ small lags, but it also produces an additional $(n_0-1)(n-n_0)$ large lags (Appendix A.6.2). The value of n_0 for which the number of small lags equals the number of large lags is (Appendix A.6.3)

$$n_0 = n + \frac{1}{2} - \frac{1}{2}\sqrt{2n^2 - 2n + 1} \simeq 0.3n.$$

This division is motivated by the fact that for estimation of (approximately) linear variograms it is useful to cover both extremes with equal proportion.

There are more small lags than large lags when $n_0 \ll 0.3n$ and vice versa when $n_0 \gg 0.3n$. To allow for the possibility of a larger proportion of small lags in the final design, we recommend a value of n_0 smaller than $0.3n$, say, $n_0 = 0.2n$, provided that an acceptable level of overall coverage can still be achieved.

Of course it must be emphasized that an optimal choice of n_0 may depend upon various other factors. If kriging is the ultimate aim of the analysis it might be helpful to determine which relative allocation (to variogram estimation and to kriging itself) gives the best overall performance of the kriging procedure (one attempt to assess this performance by simulation and its application to monitoring network design can be found in Haas (1992)). However, as indicated, this is a largely unexplored topic.

Alternative 'Direct' Approaches

The most appealing way of combining the design purposes of trend and variogram estimation would of course be an approach, which is analogous to the 'naive' technique given in the example of this chapters introduction. It is however easy to see that for correlated cases this approach is generally not applicable. Ying (1993) investigates some special cases, e.g. the covariance function

$$c(x, x'; \theta) = \theta_3 \exp\{-\theta_1 |x_1 - x_1'| - \theta_2 |x_2 - x_2'|\},$$

where $\theta \in (0, \infty)^3$ and $\eta(\cdot) \equiv 0$. For these cases he finds the estimation of the variance parameters is design independent as long as the design is asymptotically space-filling. The technique in general is of limited practical value, since Ying (1993) also demonstrates that his results are for instance not applicable to the corresponding covariance function based on the more commonly used Euclidean distances.

A very different 'direct' approach can be taken if the complete probabilistic structure of the random field model is known. An information theoretic approach based upon the Shannon entropy, which was firstly formalized by Caselton and Zidek (1984), is then possible. Their basic idea was that the uncertainty about some aspect (say parameter estimation, prediction, etc.) must be somehow reduced by the data and that a design that produces such data should minimize the overall entropy

$$\mathcal{E}[y] = E[-\log p(y) \mid \tilde{p}(y)].$$

Here, y is the collection of potential measurements and $p(\cdot)$ its probability density function, $\tilde{p}(\cdot)$ represents the state of complete ignorance and is usually chosen as an improper density $\tilde{p}(\cdot) \propto 1$.

Let us assume that the data can only be collected on a finite grid $\bar{\mathcal{X}}$, and the purpose of the design is prediction, then Shewry and Wynn (1987) have pointed out that

$$\mathcal{E}[y(\bar{\mathcal{X}})] = \mathcal{E}[y(A)] + E_{y(A)}\{\mathcal{E}[y(\bar{\mathcal{X}} \setminus A) \mid y(A)]\},$$

where $A \subseteq \bar{\mathcal{X}}$ defines a replication-free design ξ_A. Since on a finite grid the total entropy is fixed, it is the second term, the expected entropy of the data at unobserved sites conditioned on the observations, that needs to be minimized. Evidently, one can equivalently maximize the first term and thus use

$$\mathcal{J}(\xi_A) = \mathcal{E}[y(A)]$$

as a measure of information in (3.1) and this procedure is usually called maximum entropy sampling. It is widely used as a design criterion in computer simulation experiments, see Sacks et al. (1989) or Koehler and Owen (1996) for a review; an extension to the multivariate case and application to spatial sampling can be found in Bueso et al. (1999). For the numerical evaluation a number of methods same or similar to those from Section 3.3 can be applied (cf. Currin et al. (1990)), an exact algorithm is presented in Ko et al. (1995).

In the case of parameter estimation the rationale is very similar (see for instance Guttorp et al. (1993) or Sebastiani and Wynn (2000)). If it can be assumed that the joint entropy of the parameters and the data $\mathcal{E}[y(A), \beta, \theta]$ is independent of the design A, the minimization of $E_{y(A)}\{\mathcal{E}[\beta, \theta \mid y(A)]\}$ is equivalent to maximization of $\mathcal{E}[y(A)]$ and thus in these cases the two purposes lead to the same optimum designs. The maximum entropy criterion can therefore be used in a similar spirit as a compound criterion by including estimation of a parameterized covariance function, and this is indeed done by Wu and Zidek (1992). The case of nonparameterically modelling the spatial covariances is covered by Guttorp et al. (1992); an extension of the theory for detecting the impact of discrete events is provided by Schumacher and Zidek (1993).

For the special case of Gaussian linear regression the entropy is

$$\mathcal{E}[y] = \text{constant} + \log|\text{Cov}(y)|,$$

and thus we have for a random coefficient model (or a Bayesian regression model with conjugate prior)

$$\mathcal{J}(\xi_A) = |C(A)||X^T C^{-1}(A)X + D^{-1}|. \qquad (6.7)$$

In the uncorrelated case $C(\cdot) = I$ (and fixed size design) this expression coincides with D-optimality and sensitivity function (3.14) can be applied. Therefore, as pointed out by Fedorov and Müller (1988), the heuristic procedure (3.16) also makes sense for the numerical construction of maximum entropy sampling designs. Furthermore, for the correlated case and diffuse prior information, D^{-1} vanishes and we are left with a criterion that is a product of a D-criterion as treated in Chapter 5 and the maximin criterion (4.2) from Chapter 4.

6.5 Example

The construction of an air pollution monitoring network is a good example of a multiobjective design problem as Trujillo-Ventura and Ellis (1991) point out. As potential objectives they list e.g.

- the timely declaration of alert situations,
- the evaluation of the exposure of the receptors,
- the control of emissions from single sources,

- the 'learning' for long-term prognostic purposes,
- the validation of mathematical models (which is our main purpose here),
- cost considerations.

After suitable design criteria are formulated that adequately reflect those objectives, it is straightforward to use the approaches given above to accommodate any selection of objectives from this list.

A Design for Variogram Estimation

Let us return to the example of the Upper-Austrian monitoring network and the goal of efficient estimation of the random field. The efficiency of the existing network for both estimation purposes (first and second order characteristics) can be evaluated by calculating the respective values for the case of D-optimality $\Phi[M(\xi_{17}, \beta_0)] = -8.044$ and $\Phi[M'(\xi_{17}, \theta_0)] = 11.648$. To assess whether this is satisfactory close to the optimum it must be compared to the values for a respectively constructed ξ^*. The former problem was treated in Section 5.5 and for the 17-point design ξ^*_{17} in Figure 5.4 we find $\Phi[M(\xi^*_{17}, \beta_0)] = -1.892$. For the latter purpose algorithm (6.5) was run and it yielded the design ξ'^*_{17} given in Figure 6.3, which gives a criterion value $\Phi[M'(\xi'^*_{17}, \theta_0)] = 11.960$. Note that, as expected, ξ'^*_{17} has an entirely different structure than ξ^*_{17}, which produces no 'small' lags at all.

A Design for Variogram Estimation when Correlations are Ignored

By applying the techniques from section 6.3 where we effectively ignore the correlations between different lags we find a design $\xi^{\#}_{17}$ with $\Phi[M(\xi^{\#}_{17}, \beta_0)] = -8.935$ and $\Phi[M'(\xi^{\#}_{17}, \theta_0)] = 11.940$, which is expectedly better than the existing design for the purpose of variogram estimation (almost as good as ξ'^*_{17}) but slightly worse for trend estimation. A graphical presentation of the design $\xi^{\#}_{17}$, which is clustered in the center of the region, is given in Figure 6.4.

Combining Designs for Trend and Variogram Estimation

To ensure a good overall performance of the design a combined criterion $\bar{\Phi}[\cdot]$ should be optimized. Unfortunately a direct approach is not

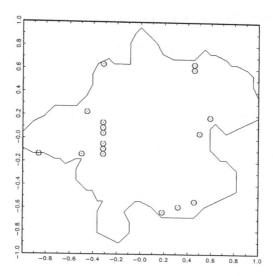

FIGURE 6.3. A D-'optimum' design for variogram estimation.

possible here, since the optimization is effectively performed over two different spaces (the site and the lag space) for the two involved criteria. An informal general recipe for that purpose is given by McBratney et al. (1981).

More formally, however, one could in the spirit of the discussion in the previous section calculate optimum designs for trend estimation with varying $n_0 \leq 17$ and then augment those designs by $17 - n_0$ points, which are optimally chosen for variogram estimation. Table 6.1 gives the values of the two design criteria for the various choices of n_0. Note that we cannot choose $n_0 < 7 = m$ without avoiding singular designs.

From these values compound criterion values can be calculated, which can be found in Figure 6.5. This graph (and also Table 6.1) reveals that at least one point (out of the 17) should be chosen to produce a small lag and that the optimal allocation should be around $n_0 = 8$, which is higher than the heuristic recommendation given above. The corresponding design, which is given in Figure 6.6, seems to be a good compromise between space-filling properties for both site- and lag-space.

It is remarkable that the existing design ξ_{17} seems to be reasonably well constructed with respect to the purpose of variogram estimation. However, the criterion value for trend estimation is very low compared with the alternative designs. This is due to the high concentration of

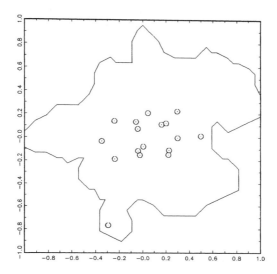

FIGURE 6.4. A design for variogram estimation when the lag correlations are ignored.

sites at one particular location, Linz. Most of the compound designs, particularly the one given in Figure 6.6, seem to be a vast improvement over ξ_{17} (which for instance has a value of $\bar{\Phi}[\xi_{17}|0.5] = 1.802$).

6.6 Exercises

1. Using the augmentation procedure construct an 'optimum' design for variogram estimation. Compare its efficiency (also with respect to trend estimation) to the other designs calculated.

2. Construct an alternative design by the multidimensional scaling technique. Report both efficiencies and computing times.

3. Use a varying portion of the 36 points for trend estimation versus variogram estimation. Give resulting designs and compare criterion values.

4. Use maximum entropy sampling to reduce the number of design points similarly to Chapter 3; compare with results therein.

5. Employ criterion (6.7) for all your designs calculated so far. Which one performs best?

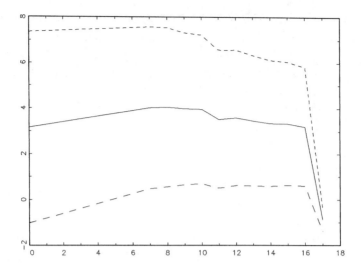

FIGURE 6.5. Compound criterion values $\bar{\Phi}[\cdot]$ vs. n_0 for $\alpha = 0.5$ (solid line), $\alpha = 0.25$ (dashed line) and $\alpha = 0.75$ (closely dashed line).

n_0	$\ln|M'(\theta_0)|$	$\ln|M(\beta_0)|$
0	11.960	-5.220
⋮	⋮	⋮
7	11.124	-3.072
8	11.042	-2.934
9	10.656	-2.681
10	10.498	-2.566
11	9.592	-2.539
12	9.550	-2.343
13	9.206	-2.264
14	8.896	-2.182
15	8.762	-2.073
16	8.433	-2.014
17	0.153	-1.892

TABLE 6.1. Design criteria for 'compound' optimum designs (here $n_0 = 0$ corresponds to ξ'^*).

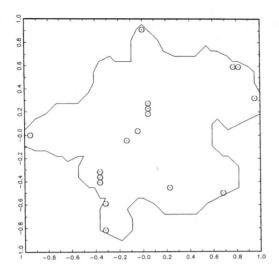

FIGURE 6.6. A 'compound' optimum design (with $n_0 = 8$).

6. Multipurpose Designs Including Designs for Variogram Fitting

Connections to Other Chapters

In this chapter several approaches for designing with the purpose of estimating the second order characteristics were presented. Finally methods from all chapters (especially 5 and 6) were employed to construct designs with a good overall performance.

References

Abt, M. and Welch, W.J. (1998). Fisher information and maximum likelihood estimation of covariance parameters in Gaussian stochastic processes. *The Canadian Journal of Statistics*, 26(1):127–137.

Atkinson, A.C. and Cook, R.D. (1995). D-Optimum designs for heteroscedastic linear models. *Journal of the American Statistical Association*, 90(429):204–212.

Bogaert, P. and Russo, D. (1999). Optimal spatial sampling design for the estimation of the variogram based on a least squares approach. *Water Resources Research*, 35(4):1275–1289.

Brimkulov, U.N., Krug, G.K. and Savanov, V.L. (1980). Numerical construction of exact experimental designs when the measurements are correlated (in Russian). *Zavodskaya Laboratoria (Industrial Laboratory)*, 36:435–442.

Bueso, M.C., Angulo, J.M., Cruz-Sanjulian, J., and Garcia-Arostegui, J.L. (1999). Optimal spatial sampling design in a multivariate framework. *Mathematical Geology*, 31(5):507–525.

Caselton, W.F. and Zidek, J.V. (1984). Optimal monitoring networks. *Statistics & Probability Letters*, 2:223–227.

Cook, D. and Fedorov, V. (1995). Constrained optimization of experimental design (with discussion). *Statistics*, 26:129–178.

Cook, R.D. and Nachtsheim, C.J. (1982). Model robust, linear-optimal designs. *Technometrics*, 24:49–54.

Cook, R.D. and Wong, W.K. (1994). On the equivalence of constrained and compound optimal designs. *Journal of the American Statistical Association*, 89:687–692.

Cox, T.F. and Cox, M.A.A. (1994). *Multidimensional Scaling*. Chapman & Hall, London.

Cressie, N. (1985). Fitting variogram models by weighted least squares. *Mathematical Geology*, 17:563–586.

Currin, C., Mitchell, T., Morris, M. and Ylvisaker, D. (1990). Bayesian predictions of deterministic functions, with applications to the design and analysis of computer experiments. *Journal of the American Statistical Association*, 86(416):953–963.

Dette, H. (1990). A generalization of D- and D_1-optimal designs in polynomial regression. *The Annals of Statistics*, 18(4):1784–1804.

Fedorov, V.V. and Flanagan, D. (1997). Optimal monitoring network design based on Mercer's expansion of covariance kernel. *Journal of Combinatorics, Information & System Sciences*, 23:237–250.

Fedorov, V.V. and Müller, W.G. (1988). Two approaches in optimization of observing networks. In Dodge, Y., Fedorov, V.V. and Wynn, H.P., editors, *Optimal Design and Analysis of Experiments*, North-Holland, 239–256.

Fedorov, V.V. (1972). *Theory of Optimal Experiments*. Academic Press, New York.

Guttorp, P., Le, N.D., Sampson, P.D. and Zidek, J.V. (1993). Using entropy in the redesign of an environmental monitoring network. In Patil, G.P., and Rao, C.R., editors, *Multivariate Environmental Statistics*, Elsevier Science, Amsterdam, 175–202.

Guttorp, P., Sampson, P.D., and Newman, K. (1992). Nonparametric estimation of spatial covariance with application to monitoring network evaluation. In Walden, A.T. and Guttorp, P., editors, *Statistics in the Environmental and Earth Sciences. New Developments in Theory and Practice*, Edward Arnold, London—Baltimore, 39–51.

Haas, T.C. (1992). Redesigning continental-scale monitoring networks. *Atmospheric Environment*, 26A(18):3323–3353.

Ko, C.W., Lee, J. and Queyranne, M. (1995). An exact algorithm for maximum entropy sampling. *Operations Research*, 43(3):684–691.

Koehler, J.R. and Owen, A.B. (1996). Computer experiments. In Gosh, S. and Rao, C.R., editors, *Handbook of Statistics*, volume 13. Elsevier, Amsterdam, 261–308.

Lamorey, G. and Jacobson, E. (1995). Estimation of semivariogram parameters and evaluation of the effects of data sparsity. *Mathematical Geology*, 27(3):327–358.

Läuter, E. (1976). Optimal multipurpose designs for regression models. *Mathematische Operationsforschung und Statistik, Series Statistics*, 7:51–68.

Lee, S.Y. (1984). Multidimensional scaling models with inequality and equality constraints. *Communications in Statistics, Simulations and Computations*, 13(1):127–140.

Maljutov, M.B. (1988). Design and analysis in generalized regression model F. In Fedorov, V.V. and Läuter, H., editors, *Lecture Notes in Economical and Mathematical Systems*. Springer Verlag, Berlin, 297:72–76.

McBratney, A.B., Webster, R. and Burgess, T.M. (1981). The design of optimal sampling schemes for local estimation and mapping of regionalized variables — I. *Computers and Geosciences*, 7:331–334.

Morris, M.D. (1991). On counting the number of data pairs for semivariogram estimation. *Mathematical Geology*, 23(7):929–943.

Müller, W.G. and Zimmerman, D.L. (1995). An algorithm for sampling optimization for semivariogram estimation. In Kitsos, C.P. and Müller, W.G., editors, *Model-Oriented Data Analysis 4*, Heidelberg. Physica, 173–178.

Müller, W.G. and Zimmerman, D.L. (1999). Optimal design for variogram estimation. *Environmetrics*, 10:23–37.

Oehlert, G.W. (1997). Shrinking a wet deposition network. *Atmospheric Environment*, 30(8):1347–1357.

Pettitt, A.N. and McBratney, A.B. (1993). Sampling designs for estimating spatial variance components. *Applied Statistics*, 42(1):185–209.

Prince and the Revolution. (1986). Kiss. In *Parade*. Warner Bros. Records Inc.

Russo, D. (1984). Design of an optimal sampling network for estimating the variogram. *Soil Science Society of America Journal*, 48:708–716.

Sacks, J., Welch, W.J., Mitchell, T.J. and Wynn, H.P. (1989). Design and analysis of computer experiments. *Statistical Science*, 4(4):409–435.

Sacks, J. and Ylvisaker, D. (1966). Design for regression problems with correlated errors. *Annals of Mathematical Statistics*, 37:66–89.

Schumacher, P. and Zidek, J.V. (1993). Using prior information in designing intervention detection experiments. *The Annals of Statistics*, 21(1):447–463.

Sebastiani, P. and Wynn, H.P. (2000). Maximum entropy sampling and optimal Bayesian experimental design. *Journal of the Royal Statistical Society, Series B*, 62(1):145–157.

Shewry, M.C. and Wynn, H.P. (1987). Maximum entropy sampling. *Journal of Applied Statistics*, 14(2):165–170.

Stein, M.L. (1988). Asymptotically efficient prediction of a random field with misspecified covariance function. *The Annals of Statistics*, 16(1):55–63.

Steinberg, D.M., Rabinowitz, N., Shimshoni, Y., and Mizrachi, D. (1995). Configuring a seismographic network for optimal monitoring of fault lines and multiple sources. *Bulletin of the Seismological Society of America*, 85(6):1847–1857.

Trujillo-Ventura, A. and Ellis, J.H. (1991). Multiobjective air pollution monitoring network design. *Atmospheric Environment*, 25A(2):469–479.

Warrick, A.W. and Myers, D.E. (1987). Optimization of sampling locations for variogram calculations. *Water Resources Research*, 23:496–500.

Wu, S. and Zidek, J.V. (1992). An entropy-based analysis of data from selected NADP/NTN network sites for 1983–1986. *Atmospheric Environment*, 26A(11):2089–2103.

Ying, Z. (1993). Maximum likelihood estimation of parameters under a spatial sampling scheme. *The Annals of Statistics*, 21(3):1567–1590.

Zimmerman, D.L. and Homer, K. (1991). A network design criterion for estimating selected attributes of the semivariogram. *Environmetrics*, 2:425–441.

Appendix

A.1 Data Sets

Example Set: SO_2 in Upper Austria

The rows and columns indicate the date and station number respectively. Station numbers are as follows: 01 Linz—Hauserhof, 03 Linz—Urfahr, 04 Traun, 05 Asten, 06 Wels, 07 Vöcklabruck, 08 Perg, 09 Steyr, 10 Braunau, 12 Linz—Kleinmünchen, 13 Linz—Ursulinenhof, 14 Linz—ORF-Zentrum, 15 Linz—25er-Turm, 16 Linz—Berufsschulzentrum, 17 Steyregg-Weih, 18 Lenzing, 20 Schöneben. The measurements are daily averages of SO_2 concentrations in units of mg/m^3.

date	01	03	04	05	06	07	08	09	10	12	13	14	15	16	17	18	20
94-																	
01-26	1	0	1	3	6	4	4	2	0	0	3	2	0	1	2	1	0
02-02	5	6	9	8	15	10	7	6	9	10	6	7	5	8	13	5	1
02-09	20	20	19	18	26	10	26	14	8	29	20	23	26	20	22	12	18
02-12	19	21	13	16	15	12	18	15	16	23	20	22	23	25	25	22	22
02-16	11	10	9	15	18	19	14	10	11	14	17	11	12	41	12	47	22
02-17	14	15	15	15	13	18	15	8	10	17	19	16	14	27	19	15	14
02-18	30	16	12	13	14	16	13	12	10	17	26	28	13	18	16	22	9
02-19	46	44	46	42	35	27	33	20	18	57	54	53	52	53	54	34	33
02-20	50	34	42	33	32	22	33	20	13	43	50	47	33	42	41	20	3
02-21	47	18	24	21	18	22	19	13	13	25	29	30	24	25	28	16	4
02-22	22	12	15	12	10	14	8	7	4	12	19	17	14	14	13	11	0
02-23	11	12	10	10	10	11	5	7	5	11	15	11	14	10	22	10	0
02-25	3	3	3	4	5	7	7	4	0	3	5	2	5	3	5	2	0
02-26	21	4	5	4	2	3	2	3	1	0	12	15	4	4	4	14	0
02-27	22	8	9	6	4	4	4	5	2	7	13	16	9	16	9	14	1
02-28	17	13	11	11	9	5	7	5	4	14	22	20	17	13	16	11	1
03-01	17	16	9	6	10	5	6	5	1	9	23	17	18	12	8	5	2
⋮	⋮	⋮	⋮	⋮	⋮	⋮	⋮	⋮	⋮	⋮	⋮	⋮	⋮	⋮	⋮	⋮	⋮
95-																	
10-31	8	7	8	6	9	3	6	11	6	7	8	8	5	8	4	10	8
11-01	4	4	2	3	3	3	2	1	2	2	3	3	1	5	3	1	1
11-02	2	1	5	2	2	3	3	2	1	1	4	2	1	1	3	0	2
11-03	0	0	4	0	2	1	0	1	0	1	3	0	0	0	1	1	1
11-11	17	6	9	8	8	2	10	3	1	8	12	7	4	20	8	9	8
11-18	3	3	4	5	4	2	4	2	1	4	6	4	4	4	2	3	6
11-21	6	7	11	6	11	9	8	6	6	7	8	8	7	21	11	11	7
11-22	26	6	11	5	12	11	6	6	6	18	14	5	8	14	11	1
12-06	27	21	27	27	30	19	26	20	16	26	14	18	18	56	31	21	26
12-07	26	16	21	17	20	14	15	16	13	20	9	21	13	24	29	13	12
12-08	28	12	17	15	17	9	12	15	8	15	4	20	9	20	17	22	6
12-09	10	9	13	9	8	2	6	6	3	9	6	8	7	12	12	6	5
12-10	9	8	10	6	2	1	3	2	1	6	5	9	5	7	15	10	3
12-11	4	6	5	3	8	6	3	7	5	4	2	6	4	4	11	3	10
12-22	15	8	6	4	5	5	2	8	2	3	12	18	4	7	1	7	0

Exercise Set: Chlorid in Südliches Tullnerfeld

The rows and columns indicate the date and station number respectively. The measurements are daily averages of Cl concentrations in units of mg/l.

date	411	429	849	854	...	...	...	5320	5321	5322	5323
16.01.92					...	...	...				
20.01.92					...	...	...				
21.01.92	22.5				...	...	...				
22.01.92				73.2	...	...	...				
23.01.92		57.5	60.1		...	...	...				
27.01.92					...	...	...				
10.02.92		60			...	...	...				
18.02.92	24.3				...	...	...				
16.03.92					...	...	...				
17.03.92	21		57		...	...	...				
18.03.92					...	...	...				
19.03.92		55		64	...	...	...				
25.03.92					...	...	...				
07.04.92					...	...	...				
13.05.92	22				...	...	...				
27.05.92		55			...	...	...				
10.06.92					...	...	...				
22.06.92	20.9				...	...	...				
23.06.92					...	...	...				
24.06.92			58.1	65.3	...	...	...				
25.06.92		59			...	...	...				
03.08.92					...	...	...				
10.08.92				59	...	...	...				
11.08.92		60			...	...	...				
31.08.92	18.6				...	...	...				
01.09.92					...	...	...				
02.09.92			53.8	64.8	...	...	...				
03.09.92	22.5				...	...	...				
08.09.92					...	...	...				
09.09.92		56			...	...	...				
28.09.92			60		...	...	...				
17.11.92	20				...	...	...				
⋮	⋮	⋮	⋮	⋮	⋮	⋮	⋮	⋮	⋮	⋮	⋮
09.09.96			60		...	...	...				
10.09.96					...	...	...				
11.09.96					...	...	...				
12.09.96	13.3				...	...	...	114			
16.09.96					...	...	...			55.3	60.6
17.09.96		57.3	60	69.5	...	...	...				
14.10.96	13				...	...	...				
16.10.96			58		...	...	...				
20.11.96					...	...	...				
03.12.96			57.35	82.7	...	...	...				
05.12.96	13.1				...	...	...	117	119	28.5	59.9
08.12.96					...	...	...				
17.12.96			58.1		...	...	...				
05.02.97	14.3				...	...	...				
06.02.97			59		...	...	...	117			
18.03.97				78.9	...	...	...				
19.03.97					...	...	...				46.2
20.03.97		57	56.8		...	...	...				
24.03.97			56		...	...	...			60.1	
25.03.97	14.1				...	...	...				

A.2 Proofs for Chapter 2

A.2.1 Inconsistency of the W.L.S. Variogram Estimator (Müller (1999))

For a particular sampling scheme (like the one given in Müller (1999), which is created by the algorithm of Müller and Zimmerman (1995)) the conjecture of inconsistency of (2.23) becomes very transparent. Under the assumption of negligible off-diagonal elements of $\text{Cov}(\hat{\gamma})$ it is easy to adopt the proof of inconsistency from Fedorov (1974) to find a simple correction of (2.23): We can split the 'observations' $\hat{\gamma}_k$ into $\gamma(h_k, \theta^+) + \varepsilon_k$, where θ^+ denotes the true value of the parameter. The sum to be minimized from (2.23) can in the limit ($K \to \infty$) also be separated into

$$\sum_{k=1}^{\infty} \gamma^{-2}(h_k, \theta)[\gamma(h_k, \theta^+) - \gamma(h_k, \theta)]^2 + \sum_{k=1}^{\infty} \gamma^{-2}(h_k, \theta)\varepsilon_k^2.$$

Due to the law of large numbers (and assuming $E(\varepsilon_k) = 0$) this gives

$$\sum_{k=1}^{\infty} \left[\frac{\gamma(h_k, \theta^+)}{\gamma(h_k, \theta)} - 1\right]^2 + \sum_{k=1}^{\infty} \frac{2\gamma^2(h_k, \theta^+)}{\gamma^2(h_k, \theta)},$$

and after multiplying out and regrouping we find that it is equivalent to minimize

$$\sum_{k=1}^{\infty} \left[\frac{\gamma(h_k, \theta^+)}{\gamma(h_k, \theta)} - \frac{1}{3}\right]^2,$$

which clearly gives under mild conditions on γ (see e.g. Pázman (1993) for consistency conditions for nonlinear regression estimates) that the W.L.S. estimator $\gamma(h, \tilde{\theta})$ consistently estimates $3\gamma(h, \theta^+)$ rather than $\gamma(h, \theta^+)$. □

A.2.2 Locality of Local Regression Estimates (Müller (1991))

Ripley (1981) has pointed out that $\hat{y}(x_0)$ from a local regression should not depend upon the size of the region $\mathcal{X}$ where the data is collected. We want the predictor to be a 'true' local average in this sense. Thus, when the weight function is not vanishing we will require it to decay fast enough to have this property. The condition given in Section 2.3 is sufficient for this purpose as can be seen from a general proof for dimensionality $\bar{d}$ of the design space.

For a homogeneous arrangement of data points (design) it is assumed that there are about $\text{const} \times \bar{h}^{\bar{d}-1}\Delta\bar{h}$ data points within a distance $\bar{h}$ to $\bar{h} + \Delta\bar{h}$ from a point $x_0 \in \mathcal{X}$, since the volume of a $\bar{d}$-dimensional sphere of radius $\bar{h}d$ is proportional to $\bar{h}^{\bar{d}}$. Hence the volume of the given area is proportional to $[(\bar{h} + \Delta\bar{h})^{\bar{d}} - \bar{h}^{\bar{d}}]$. Since the first term of the left hand polynomial $\bar{h}^{\bar{d}}$ cancels out, its dominating part is the second term $\bar{h}^{\bar{d}-1}\Delta\bar{h}$.

These points contribute about $\text{const} \times \bar{h}^{\bar{d}-1}\Delta\bar{h}\lambda(\bar{h})$ to the total of weights. Thus unless the integral $\int_{\bar{c}}^{\infty} \bar{h}^{\bar{d}-1}\lambda(\bar{h})d\bar{h}$, with $\bar{c} > 0$ is not finite, $\hat{y}(x_0)$ entirely depends upon how large $\mathcal{X}$ is chosen.

But if now $\lambda(\bar{h}) = o(\bar{h}^{-\bar{d}})$ is guaranteed as $\bar{h} \to \infty$, then if for instance $\lambda(\bar{h}) = \bar{h}^{-\bar{d}-\alpha}$ with $\alpha > 0$ and $\bar{c} = 1$ it follows:

$$\int_1^{\infty} \bar{h}^{\bar{d}-1}\bar{h}^{-\bar{d}-\alpha}d\bar{h} = \int_1^{\infty} \bar{h}^{-1-\alpha}d\bar{h} = -\frac{\bar{h}^{-\alpha}}{\alpha}\Big|_1^{\infty} = -\frac{1}{\alpha\bar{h}^{\alpha}}\Big|_1^{\infty} = \frac{1}{\alpha} < \infty$$

$\square$

A.3 Proofs for Chapter 3

A.3.1 Equivalence of Designs for Localizing Extremal Values with Standard Polynomial Designs (Fedorov and Müller (1997))

The problem of searching the minimum (or maximum) of (3.12) is only reasonable in the case of a nonzero discriminant, i.e. $4\tau_{11}\tau_{22} - \tau_{12}^2 \neq 0$, which will be assumed in what follows.

$$\frac{\partial E[y(x,w)]}{\partial w} = \begin{pmatrix} 2\tau_{11}(\tau_1 - x_{[1]}) + \tau_{12}(\tau_2 - x_{[2]}) \\ 2\tau_{22}(\tau_2 - x_{[2]}) + \tau_{12}(\tau_1 - x_{[1]}) \\ 1 \\ (\tau_1 - x_{[1]})^2 \\ (\tau_2 - x_{[2]})^2 \\ (\tau_1 - x_{[1]})(\tau_2 - x_{[2]}) \end{pmatrix} \quad (A.8)$$

$$= \underbrace{\begin{pmatrix} B_{11} & 0 \\ 0 & I \end{pmatrix}}_{B} \times \underbrace{\begin{pmatrix} u_1 \\ u_2 \\ 1 \\ u_1^2 \\ u_2^2 \\ u_1 u_2 \end{pmatrix}}_{f(u)},$$

where $u_1 = \tau_1 - x_{[1]}$, $u_2 = \tau_2 - x_{[2]}$, $w^T = (\tau_1, \tau_2, \tau_0, \tau_{11}, \tau_{22}, \tau_{12})$, and the 2×2 upper-left submatrix of B contains

$$B_{11} = \begin{pmatrix} 2\tau_{11} & \tau_{12} \\ \tau_{12} & 2\tau_{22} \end{pmatrix}.$$

It follows from (A.8) that the information matrix for a single observation made at point x equals

$$m(x) = Bf(u)f^T(u)B,$$

and subsequently for the design ξ_u with support S_{ξ_u}

$$M(\xi) = B \sum_{S_{\xi_u}} f(u)f^T(u)\xi_u(u)B = B\mathcal{M}(\xi_u)B,$$

where $\mathcal{U} = \{u : u = \tau - x, x \in \mathcal{X}\}$ and $\tau^T = (\tau_1, \tau_2)$.

As a reasonable design criterion we may choose D_s-optimality, i.e. D_L-optimality with $L = \begin{pmatrix} 1 & 0 & 0 & 0 & 0 \\ 0 & 1 & 0 & 0 & 0 \end{pmatrix}^T$.

Then we have

$$\Phi[M(\xi)] = -|M_{11}^{-1}|, \tag{A.9}$$

where M_{11}^{-1} is the respective upper-left block of M^{-1} and existence of this matrix is assumed for the sake of simplicity.

Matrix B is regular (we have assumed that $4\tau_{11}\tau_{22} - \tau_{12}^2 \neq 0$). Therefore

$$M^{-1}(\xi) = B^{-1}\mathcal{M}^{-1}(\xi_u)B^{-1} = \begin{pmatrix} B_{11}^{-1}\mathcal{M}_{11}^{-1}B_{11}^{-1} & B_{11}^{-1}\mathcal{M}_{12}^{-1} \\ \mathcal{M}_{21}^{-1}B_{11}^{-1} & \mathcal{M}_{22}^{-1} \end{pmatrix}. \tag{A.10}$$

From (A.10) it follows that

$$\Phi[M(\xi)] = -|B_{11}|^{-2}|\mathcal{M}_{11}^{-1}|,$$

and hence

$$\max_\xi \Phi[M(\xi)] = |B_{11}|^{-2} \min_{\xi_u} |\mathcal{M}_{11}^{-1}(\xi_u)|.$$

Thus, the problem of the optimum design construction for (3.5) is equivalent to

$$\xi_u^* = \arg \min_{\xi_u} |\mathcal{M}_{11}^{-1}(\xi_u)|, \tag{A.11}$$

which is a standard problem in experimental design, and thus can be numerically evaluated by one of the algorithms from section 3.3. It is important to notice that (A.11) includes only two unknown parameters τ_1 and τ_2 (cf. Hill (1980)). □

A.3.2 Equivalence of Bayesian Design with a Heuristic Procedure (Fedorov and Müller (1989))

If the data are generated by (3.13) the variance-covariance matrix of the observations is given by

$$K = E[(y_j - \dot{\eta}^T \beta)(y_j - \dot{\eta}^T \beta)^T] = \sigma^2(I + \dot{\eta}^T D \dot{\eta}).$$

By applying a standard inversion formula (cf. e.g. Dhrymes (1984)) we have

$$\sigma^2 K^{-1} = I - \dot{\eta}^T \left[\frac{\sigma^2}{N} D^{-1} + M(\xi_N) \right]^{-1} \dot{\eta},$$

and thus for a particular $x_{i^-} \in \xi_N$ we have that

$$i^- = \arg\max_i K_{ii}^{-1}$$

is equivalent to

$$i^- = \arg\min_i \phi_\beta(x_i, \xi_N).$$

□

A.4 Proofs for Chapter 4

A.4.1. A Lower Bound for 'Coffee-house' Designs (Müller (2000))

In the one-dimensional setup it is evident that the regular equispaced sequence of points provides the maximin distance designs and we therefore have

$$h^*_{\min} = \frac{1}{n-1}.$$

The corresponding minimum distance in coffee-house designs is

$$h_{\min} = \max_{i \in \mathbb{N}} \frac{1}{2^i} \quad \text{s.t. } 2^i \geq n - 1,$$

and the ratio is thus given by

$$\frac{h_{\min}}{h^*_{\min}} \geq \frac{2^{i-1} - 1}{2^i}.$$

The above equality holds at $n = 2^{i-1} + 2$ and therefore by resubstituting n in the second term we have as a lower bound

$$\frac{h_{\min}}{h^*_{\min}} \geq \frac{n-1}{2(n-2)} > \frac{1}{2}.$$

□

A.4.2. Variance of a Local Regression Estimate

Due to linearity of $\hat{\beta}(_jx)$ (in a local polynomial model), its variance-covariance matrix is given by

$$L(_jx)\mathrm{Cov}(y)L^T(_jx) = (X_j^T V_j^{-1} X_j)^{-1} X_j^T V_j^{-1} \mathrm{Cov}(y) V_j^{-1} X_j \times$$
$$\times (X_j^T V_j^{-1} X_j)^{-1}.$$

Hence under the assumption of independence and $\sigma^2(\cdot) \propto \frac{1}{w(\cdot)}$ we have

$$\mathrm{Cov}(\hat{\beta}(_jx)) = (X_j^T V_j^{-1} X_j)^{-1},$$

where X_j contains the regressors from (2.14) at $x_0 = {_jx}$, and

$$[V_j]_{ii'} = \begin{cases} w^{-1}(h_i) & i = i' \\ 0 & i \neq i' \end{cases}.$$

Extracting the upper left element yields

$$\mathrm{Var}(\hat{\beta}_0(_jx)) = \mathrm{tr} O (X_j^T V_j^{-1} X_j)^{-1},$$

which due to (2.13) is equivalent to (4.4). □

A.4.3. The Sensitivity Function for Local Regression Designs (Müller (1991))

For the sake of simplicity we turn again from exact designs ξ_N to approximate designs ξ, which impose a measure on the space spanned by the regressors. It is clear that if ξ^* is the optimum design and ξ' is any other, that due to the linearity of $\mathcal{J}(\xi_N)$, cf. (4.5), the following inequality holds at point $\xi = \xi^*$:

$$\frac{\partial \mathcal{J}(\xi)}{\partial \alpha} \leq 0, \tag{A.12}$$

if $\xi = (1-\alpha)\xi^* + \alpha\xi'$, with $M_j(\xi) = (1-\alpha)M_j(\xi^*) + \alpha M_j(\xi')$ (see for similar cases Section 3.2).

Differentiating (4.5) yields

$$\frac{\partial \mathcal{J}(\xi)}{\partial \alpha} = -\frac{\partial \operatorname{tr} \sum_{j=1}^q W_j M_j^{-1}(\xi)}{\partial \alpha}$$

$$= -\operatorname{tr} \sum_{j=1}^q W_j \frac{\partial M_j^{-1}(\xi)}{\partial \alpha} \leq 0,$$

and following Corollary (41) of Dhrymes (1984)

$$= \operatorname{tr} \sum_{j=1}^q W_j M_j^{-1}(\xi) \frac{\partial M_j(\xi)}{\partial \alpha} M_j^{-1}(\xi) \leq 0,$$

after inserting $M_j(\xi) = (1-\alpha)M_j(\xi^*) + \alpha M_j(\xi')$:

$$\operatorname{tr} \sum_{j=1}^q W_j M_j^{-1}(\xi) \frac{\partial [(1-\alpha)M_j(\xi^*) + \alpha M_j(\xi')]}{\partial \alpha} M_j(\xi)^{-1}$$

$$= \operatorname{tr} \sum_{j=1}^q W_j M_j^{-1}(\xi)[M_j(\xi') - M_j(\xi^*)]M_j^{-1}(\xi) \leq 0.$$

If one now lets $\alpha \to 0$ and thus $\xi \to \xi^*$ then

$$\operatorname{tr} \sum_{j=1}^q W_j M_j^{-1}(\xi^*) M_j(\xi') M_j^{-1}(\xi^*) \geq \operatorname{tr} \sum_{j=1}^q W_j M_j^{-1}(\xi^*)$$

By inserting $M_j(\xi') = w(h_{ij}) X_{ij}{}^T X_{ij}$ for a one-point correction (i.e. $S_{\xi'} = x_i$) and by reordering we have (4.6). $\square$

A.4.4. T-Optimality in the Linear (Uncorrelated) Case

In this setup we have

$$\Delta(\xi_N) = \min_{\beta \in \mathbb{R}^m} \sum_{x \in S_{\xi_N}} \{\dot{\eta}_+^T(x)\beta_+ - \dot{\eta}^T(x)\beta\}^2 \xi(x).$$

By direct differentiation with respect to β we find

$$2 \sum_{x \in S_{\xi_n}} \dot{\eta}(x)\{\dot{\eta}_+^T(x)\beta_+ - \dot{\eta}^T(x)\beta\}\xi(x) = 0,$$

$$\{\sum_{x\in S_{\xi_N}} \dot\eta(x)\dot\eta_+^T(x)\xi(x)\}\beta_+ = \{\sum_{x\in S_{\xi_N}} \dot\eta(x)\dot\eta^T(x)\xi(x)\}\beta,$$

and thus in the nonsingular case

$$\beta = (X^T X)^{-1} X^T X_+ \beta_+.$$

The maximizing function can now be rewritten as

$$\Delta(\xi_N) = \tfrac{1}{N} \; (X^T\beta - X_+^T X(X^T X)^{-1} X^T X_+\beta_+)^T \times \\ (X^T\beta - X_+^T X(X^T X)^{-1} X^T X_+\beta_+),$$

and we have after multiplying out and regrouping

$$\Delta(\xi_N) = \frac{1}{N}\beta_+^T\{X_+^T X_+ - X_+^T X(X^T X)^{-1} X^T X_+\}\beta_+.$$

□

A.5 Proofs for Chapter 5

In the following two assumptions are made that are required for the subsequent derivations:

- Firstly it is supposed that the so-called inclusion condition holds, so that $\dot\eta(x)$ is in the column space of the matrix $C(A)$ for every $x \in A$ (cf. Pukelsheim (1993)). This is evident for some nonsingular matrices, particularly for uncorrelated observations if $n_A \geq m$.

- Secondly, $\bar{\mathcal{X}}$ must be such that $C(\bar{\mathcal{X}})$ is regular.

A.5.1 Justifications for Using $M(S_\xi)$
(Pázman and Müller (1998))

To avoid the inverse in (5.1) let us use the following equality (cf. Rao (1973)): If Q is a positive definite $m \times m$ matrix, and if $g \in \mathbb{R}^m$ is a vector, then

$$g^T Q^{-1} g = \sup_\zeta \frac{[g^T\zeta]^2}{[\zeta^T Q\zeta]},$$

where the supremum is over all non-zero vectors ζ in $\mathbb{R}^m$.

We shall use this equality to express $M(A)$ through the measure ξ_A. Note that every symmetric matrix B is completely specified by $v^T B v$, $\forall v \in \mathbb{R}^m$. Therefore, by taking $Q = C(A)$, and defining $g_v(x) = v^T f(x)$, we obtain that for every vector $v \in \mathbb{R}^m$

$$v^T M(A) v = \sup_{\zeta(\cdot) \in \mathcal{F}(A)} \frac{[\sum_{x \in A} g_v(x) \zeta(x)]^2}{[\sum_{x \in A} \sum_{x' \in A} \zeta(x) c(x, x') \zeta(x')]}, \quad (A.13)$$

where $\mathcal{F}(A)$ is the set of all nonzero functions on A.

We see that the inverse of $C(A)$ from the definition of $M(A)$ does not appear in (A.13). We have just two sums and we can proceed like in (3.2), where sums were substituted by means with respect to the design measure ξ. By this substitution, when applied to (A.13), we obtain

$$v^T M(A) v = \sup_{\zeta(\cdot) \in \mathcal{F}(A)} \frac{[\sum_{x \in A} g_v(x) \zeta(x) \xi_A(x)]^2}{[\sum_{x \in A} \sum_{x' \in A} \zeta(x) c(x, x') \zeta(x') \xi_A(x) \xi_A(x')]},$$

where we used the measure ξ_A corresponding to the design A. Note that $A = S_{\xi_A}$. By extending the right hand side to any design measure ξ we obtain

$$\sup_{\zeta(\cdot) \in \mathcal{F}(S_\xi)} \frac{[\sum_{x \in S_\xi} g_v(x) \zeta(x) \xi(x)]^2}{[\sum_{x \in S_\xi} \sum_{x' \in S_\xi} \zeta(x) c(x, x') \zeta(x') \xi(x) \xi(x')]}.$$

It is easy to show that this expression is equal to $v^T M(S_\xi) v$. Indeed, one can define $\zeta^*(x) = \zeta(x) \xi(x)$, take the supremum over all such $\zeta^*(\cdot)$, and compare with (A.13). So the mapping $\xi \to M(S_\xi)$ can be considered as a (first) extension of the information matrix for design measures.

Another justification for the use of the mapping $\xi \to M(S_\xi)$ is that it is a concave function. In the Loewner ordering we have the inequalities

$$\begin{aligned} M(S_{(1-\alpha)\xi + \alpha \xi'}) &= M(S_\xi \cup S_{\xi'}) \\ &\geq \max\{M(S_\xi), M(S_{\xi'})\} \\ &\geq (1-\alpha) M(S_\xi) + \alpha M(S_{\xi'}). \end{aligned}$$

□

If $\Phi[M]$ is concave in M and increasing then the mapping $\xi \to \Phi[M(S_\xi)]$ is concave, which follows easily from

$$\Phi[M(S_{(1-\alpha)\xi+\alpha\xi'})] \leq \Phi[(1-\alpha)M(S_\xi) + \alpha M(S_{\xi'})]$$
$$\leq (1-\alpha)\Phi[M(S_\xi)] + \alpha\Phi[M(S_{\xi'})].$$

□

A.5.2 Properties of Approximate Information Matrices (Pázman and Müller (1998))

For $\epsilon \to 0$ we have the following limit

$$\lim_{\epsilon \to 0} M^{(\epsilon)}(\xi) = n_{S_\xi} M(S_\xi),$$

where $M(S_\xi)$ is from expression (5.1) by substituting $A = S_\xi$. It follows from

$$\lim_{\epsilon \to 0} \left(\frac{\xi(x)}{\xi(\epsilon)}\right)^\epsilon = \begin{cases} 1 & \text{if } x \in S_\xi \\ 0 & \text{if } x \notin S_\xi \end{cases}. \quad \square$$

In particular $\lim_{\epsilon \to 0} M^{(\epsilon)}(\xi_A) = n_A M(A)$ for every exact design A. This means that both $M_1^{(\epsilon)}(\xi)$ and $M_2^{(\epsilon)}(\xi)$ are extensions of the information matrix (5.1). The approximation is improved when $\epsilon \to 0$, which is the reason for considering only small values of ϵ.

When the number of points in the design must not exceed a given number n, then we have to minimize $\Phi[M^{(\epsilon)}(\xi)]$ over the set

$$\Xi_n = \{\xi \in \Xi : n_{S_\xi} \leq n\},$$

which is nonconvex. To show that, let us consider two exact designs $A \neq B$, such that $n_A = n_B = n < n_{\bar{x}}$. Then $\xi_A \in \Xi_n$, $\xi_B \in \Xi_n$, but for any $\alpha \in (0,1)$ we have $\xi_{AB} = (1-\alpha)\xi_A + \alpha\xi_B \notin \Xi_n$, because $S_{\xi_{AB}} = A \cup B$, hence $n_{S_{\xi_{AB}}} > n$. □

Further investigation of $M_1^{(\epsilon)}(\cdot)$ shows us that not only $M_1^{(0)}(\xi) = n_{S_\xi} M(S_\xi)$, but also

$$\sqrt{n_{S_\xi}} M(S_\xi) \leq M_1^{(\frac{1}{2})}(\xi) \leq \frac{1}{\xi_{\max}} M(\xi) = n_{S_\xi} M(\xi),$$

cf. (3.2). This allows us to approximate the cases with and without replications by the common formula (5.3) differing only by the value of ϵ.

According to the definition of approximate design measures we observe at each $x \in S_\xi$, but when $\xi(x) < \xi_{\max}$ we have "fractional observations". For the product type definition of M_1 the continuity of the mapping $\xi \to M_1^{(\epsilon)}(\xi)$ is obvious. To prove the continuity of the mapping $\xi \to M_2^{(\epsilon)}(\xi)$, we let $\xi_s(\cdot)$, $s = 1, 2, \ldots$ now denote a sequence of design measures having the same support $S_{\xi_s} = \{x_1, \ldots, x_n\}$ such that it converges to a design measure $\xi(\cdot)$, i.e.

$$\lim_{s \to \infty} \xi_s(x) = \xi(x); \quad x \in \mathcal{X}.$$

Then for each $\epsilon > 0$

$$\lim_{s \to \infty} M_2^{(\epsilon)}(\xi_s) = M_2^{(\epsilon)}(\xi).$$

This can be seen from a step by step proof using the matrix property from e.g. Kubáček et al. (1995): Let $U = \begin{pmatrix} U_I & U_{II} \\ U_{II}^T & U_{III} \end{pmatrix}$ be a symmetric, positive definite matrix, then $U^{-1} = \begin{pmatrix} U_1 & U_2 \\ U_2^T & U_3 \end{pmatrix}$, with

$$\begin{aligned}
U_1 &= (U_I - U_{II} U_{III}^{-1} U_{II}^T)^{-1}, \\
U_2 &= -(U_I - U_{II} U_{III}^{-1} U_{II}^T)^{-1} U_{II} U_{III}^{-1}, \\
U_3 &= U_{III}^{-1} + U_{III}^{-1} U_{II}^T (U_I - U_{II} U_{III}^{-1} U_{II}^T)^{-1} U_{II} U_{III}^{-1}.
\end{aligned}$$

If U is positive semidefinite then each inverse can be substituted by a g-inverse.

Now denote $S_\xi = \{x_1, \ldots, x_{n_{S_\xi}}\}$, where $n_{S_\xi} \le n$. If $n_{S_\xi} = n$, the proof is evident. If $n_{S_\xi} < n$, take $U = C(S_\xi) + W^{(\epsilon)}(\xi_s)$ and denote U_I the upper left $(s-1) \times (s-1)$ block of this matrix. Then U_{III}^{-1} is a number

$$U_{III}^{-1} = \left[c(x_n, x_n) + \epsilon \ln \left(\frac{\xi^{(\epsilon)}}{\xi_s(x_n)} \right) \right]^{-1},$$

which tends to zero since $\xi_s(x_n) \to 0$. It follows that

$$\lim_{s \to \infty} [C(S_{\xi_s}) + W^{(\epsilon)}(\xi_s)]^{-1} = \begin{pmatrix} \lim_{s \to \infty} [C(S_{\xi_s} - \{x_n\}) + W^{(\epsilon)}(\xi_s)_I]^{-1} & 0 \\ 0 & 0 \end{pmatrix}.$$

We use the same procedure $n - n_{S_\xi}$ times to obtain finally

$$\lim_{n\to\infty}[C(S_{\xi_s}) + W^{(\epsilon)}(\xi_s)]^{-1} = \begin{pmatrix} [C(S_\xi) + W^{(\epsilon)}(\xi)]^{-1} & 0 \\ 0 & 0 \end{pmatrix}.$$

Note that $W^{(\epsilon)}(\xi)$ is an $n_{S_\xi} \times n_{S_\xi}$ matrix. The proof is finished by using the definition of $M_2^{(\epsilon)}(\xi)$. □

A.5.3. Directional Derivative for the Approximate Information Matrix M_1 (Müller and Pázman (1998))

Let us apply a differentiable optimality criterion $\Phi[\cdot]$ to the approximate information matrix $M_1^{(\epsilon)}(\xi)$. Take fixed design measures ξ and ξ', and consider the directional derivative of $\Phi[\cdot]$ at the point ξ and in the direction to ξ':

$$\phi_1^{(\epsilon)}(\xi,\xi') = \lim_{\alpha\to 0} \frac{\partial \Phi[M_1^{(\epsilon)}(\xi_\alpha)]}{\partial \alpha},$$

where

$$\xi_\alpha(x) = (1-\alpha)\xi(x) + \alpha\xi'(x).$$

We have

$$\phi_1^{(\epsilon)}(\xi,\xi') = \text{tr}\left\{\nabla\Phi[M_1^{(\epsilon)}(\xi)] \lim_{\alpha\to 0} \frac{\partial M_1^{(\epsilon)}(\xi_\alpha)}{\partial \alpha}\right\}, \quad (A.14)$$

cf. e.g. Golub and Van Loan (1983).

From the definition of $M_1^{(\epsilon)}(\xi)$ one obtains for $\alpha \in [0,1)$

$$\frac{\partial M_1^{(\epsilon)}(\xi_\alpha)}{\partial \alpha} = \sum_{x,x' \in S_{\xi_\alpha}} \dot{\eta}(x)\dot{\eta}^T(x') \frac{\partial}{\partial \alpha}\left[\frac{\xi_\alpha(x)}{\xi_\alpha^{(\epsilon)}} \frac{\xi_\alpha(x')}{\xi_\alpha^{(\epsilon)}}\right]^\epsilon,$$

since we consider only the case $C = I$.

By direct computation we have

$$\frac{\partial \ln \xi_\alpha(x)}{\partial \alpha} = \frac{\xi'(x) - \xi(x)}{\xi_\alpha(x)}, \quad (A.15)$$

$$\lim_{\alpha\to 0} \frac{\partial \ln \xi_\alpha(x)}{\partial \alpha} = \frac{\xi'(x)}{\xi(x)} - 1,$$

$$\frac{\partial}{\partial \alpha} \ln \xi_\alpha^{(\epsilon)} = \left[\sum_{x \in \mathcal{X}} \xi_\alpha^{\frac{1}{\epsilon}}(x)\right]^{-1} \sum_{x \in \mathcal{X}} \xi_\alpha^{\frac{1}{\epsilon}}(x) \left[\frac{\xi'(x) - \xi(x)}{\xi_\alpha(x)}\right], \quad (A.16)$$

$$\lim_{\alpha \to 0} \frac{\partial}{\partial \alpha} \ln \xi_\alpha^{(\epsilon)} = E_\xi^{(\epsilon)} \left[\frac{\xi'(\cdot)}{\xi(\cdot)} - 1\right],$$

$$\lim_{\epsilon \to 0} \lim_{\alpha \to 0} \frac{\partial}{\partial \alpha} \ln \xi_\alpha^{(\epsilon)} = E_{B_\xi} \left[\frac{\xi'(\cdot)}{\xi_{\max}} - 1\right],$$

where we defined for any function $h(x)$

$$E_\xi^{(\epsilon)}[h(\cdot)] = [\sum_{x \in \mathcal{X}} \xi^{\frac{1}{\epsilon}}(x)]^{-1} \sum_{x \in \mathcal{X}} \xi^{\frac{1}{\epsilon}}(x) h(x),$$

and

$$E_{B_\xi}[h(\cdot)] = \frac{1}{n_{B_\xi}} \sum_{x \in B_\xi} h(x).$$

To simplify the differentiation it is better to write

$$\left[\frac{\xi_\alpha(x)}{\xi_\alpha^{(\epsilon)}}\right]^\epsilon = \exp\{\epsilon[\ln \xi_\alpha(x) - \ln \xi_\alpha^{(\epsilon)}]\}$$

which allows to use (A.15) and (A.17) directly.

In the following we will consider only the case that $S_{\xi'} \subset S_\xi \subset \bar{\mathcal{X}}$, which is sufficient for the construction of the algorithm. We then have (by substituting $\epsilon := 2\epsilon$)

$$\lim_{\epsilon \to 0} \frac{1}{\epsilon} \frac{\partial M_1^{(\epsilon)}(\xi_\alpha)}{\partial \alpha} = \sum_{x \in S_\xi} \left(\frac{\xi(x)}{\xi^{(\epsilon)}}\right)^\epsilon \dot{\eta}(x) \dot{\eta}^T(x) \left[\frac{\xi'(x)}{\xi(x)} - 1\right]$$

$$- E_\xi^{(\epsilon)} \left[\frac{\xi'(\cdot)}{\xi(\cdot)} - 1\right] M_1^{(\epsilon)}(\xi). \quad (A.17)$$

In the particular case that $\xi'(x)$ is a one-point design supported on $x^* \in S_\xi$ we then obtain

$$\lim_{\epsilon \to 0} \frac{1}{\epsilon} \frac{\partial M_1^{(\epsilon)}(\xi_\alpha)}{\partial \alpha} = \left(\frac{\xi(x^*)}{\xi^{(\epsilon)}}\right)^\epsilon \frac{\dot{\eta}(x^*) \dot{\eta}^T(x^*)}{\xi(x^*)} - E_\xi^{(\epsilon)} \left[\frac{\mathcal{I}_{x^*}(\cdot)}{\xi(\cdot)}\right] M_1^{(\epsilon)}(\xi). \quad (A.18)$$

Thus after combining (A.18) with (A.14) we finally arrive at

$$\tilde{\phi}_1(x^*, \xi) = \lim_{\epsilon \to 0} \frac{1}{\epsilon} \phi_1^{(\epsilon)}(\xi, x^*) = \left\{\frac{r(x^*)}{\xi(x^*)} - \frac{\mathcal{I}_{B_\xi}(x^*)}{\xi_{\max} n_{B_\xi}} q(\xi)\right\},$$

where
$$q(\xi) = \mathrm{tr}\nabla\Phi[M(S_\xi)]M(S_\xi),$$
and
$$r(x^*) = \dot{\eta}^T(x^*)\nabla\Phi[M(S_\xi)]\dot{\eta}(x^*).$$

A.5.4. Directional Derivative of the Approximate Information Matrix M_2 (Müller and Pázman (1996))

Similar to the previous subsection we can use (5.6) as the starting point of the derivations. For $\alpha > 0$ we have from (5.10) and the modification by ρ

$$\frac{\partial M_{2,\kappa}^{(\epsilon)}(\xi_\alpha)}{\partial \alpha} = -\rho \sum_{x,x',u \in S_{\xi_\alpha}} \dot{\eta}(x)[C(S_{\xi_\alpha}) + \rho W_\kappa^{(\epsilon)}(\xi_\alpha)]_{x,u}^{-1}$$

$$\times \left\{ \frac{\partial}{\partial \alpha} \ln \left[\frac{\epsilon\xi_\alpha}{\epsilon\xi_\alpha(u)} \right]^\epsilon \right\} [C(S_{\xi_\alpha}) + \rho W_\kappa^{(\epsilon)}(\xi_\alpha)]_{u,x'}^{-1} \dot{\eta}^T(x'),$$

where again $\xi_\alpha = (1-\alpha)\xi + \alpha\xi'$, and $S_{\xi_\alpha} = S_\xi \cup S_{\xi'}$ does not depend on $\alpha > 0$.

So the derivative of the criterion function based on the information matrix $M_{2,\kappa}^{(\epsilon)}(\xi)$ is

$$\frac{\partial}{\partial \alpha}\Phi[M_{2,\kappa}^{(\epsilon)}(\xi_\alpha)] = -\rho \sum_{u \in S_{\xi_\alpha}} g_\epsilon(\xi_\alpha, u) \frac{\partial}{\partial \alpha} \ln\left[\frac{\epsilon\xi_\alpha}{\epsilon\xi_\alpha(u)}\right]^\epsilon,$$

where
$$g_\epsilon(\xi, u) = a_\epsilon^T(u)\nabla\Phi[M_{2,\kappa}^{(\epsilon)}(\xi)]a_\epsilon(u) \qquad (A.19)$$
and
$$a_\epsilon(u) = \sum_{x \in S_\xi} [C(S_\xi) + \rho W_\kappa^{(\epsilon)}(\xi)]_{u,x}^{-1} \dot{\eta}(x).$$

By taking limits we obtain

$$\lim_{\epsilon \to 0} \lim_{\alpha \to 0} \frac{\partial}{\partial \alpha}\Phi[M_{2,\kappa}^{(\epsilon)}(\xi_\alpha)] = \qquad (A.20)$$

$$-\rho \sum_{x \in \mathcal{X}} \lim_{\epsilon \to 0} g_\epsilon(\xi, x) \lim_{\epsilon \to 0} \lim_{\alpha \to 0} \left[\frac{\partial}{\partial \alpha}\epsilon \ln {}_\epsilon\xi_\alpha - \frac{\partial}{\partial \alpha}\epsilon \ln {}_\epsilon\xi_\alpha(x)\right].$$

The first term of this limit can be found by simply replacing $M_2^{(\epsilon)}$ by its limit. We have

$$\lim_{\alpha \to 0} \ln \left(\frac{\epsilon \xi_\alpha}{\epsilon \xi_\alpha(x)} \right) = \begin{cases} \ln \frac{\epsilon \xi}{\epsilon \xi(x)}; & x \in S_\xi \\ \infty & x \in S_{\xi'} \setminus S_\xi \end{cases}.$$

Hence, we obtain, like in the proof above

$$\lim_{\alpha \to 0} [C(S_{\xi_\alpha}) + \rho W_\kappa^{(\epsilon)}(\xi_\alpha)]^{-1} = \begin{pmatrix} [C(S_\xi) + \rho W_\kappa^{(\epsilon)}(\xi)]^{-1} & 0 \\ 0 & 0 \end{pmatrix},$$

$$\lim_{\epsilon \to 0} \lim_{\alpha \to 0} [C(S_{\xi_\alpha}) + \rho W_\kappa^{(\epsilon)}(\xi_\alpha)]^{-1} = \begin{pmatrix} [C(S_\xi) + \rho V_\kappa^{(\epsilon)}(\xi)]^{-1} & 0 \\ 0 & 0 \end{pmatrix},$$

and thus

$$\lim_{\epsilon \to 0} M_{2,\kappa}^{(\epsilon)}(\xi) = \sum_{x,x' \in S_\xi} \dot{\eta}(x)[C(S_\xi) + \rho V_\kappa(\xi)]^{-1} \dot{\eta}^T(x'),$$

where $V_\kappa(\xi)$ is a diagonal matrix with entries

$$[V_\kappa(\xi)]_{x,x} = \begin{cases} \ln \frac{\xi_{\max}}{\xi(x)}; & \text{if } \xi_{\max} < \kappa \\ \ln \frac{\kappa}{\xi(x)}; & \text{if } \xi(x) < \kappa < \xi_{\max} \\ 0; & \text{if } \xi(x) > \kappa, \end{cases}$$

which is just

$$\lim_{\epsilon \to 0} W_\kappa^{(\epsilon)}(\xi) = V_\kappa(\xi),$$

as shown below. Note that $[V_\kappa(\xi)]_{x,x}$ can be continuously extended to the cases $\kappa = \xi(x)$ and $\kappa = \xi_{\max}$. For finding the entries of $V_\kappa(\cdot)$ we have to investigate the limit behavior of the terms

$$\ln \left(\frac{\epsilon \xi}{\epsilon \xi(x)} \right)^\epsilon = \epsilon \ln[\epsilon \xi] - \epsilon \ln[\epsilon \xi(x)].$$

It turns out that we have to distinguish the cases:

- First, if $\xi(x) > \kappa$, then (5.11) implies $\lim_{\epsilon \to 0} [\epsilon \xi(x)] = \xi(x) - \kappa$, hence $\lim_{\epsilon \to 0} \epsilon \ln[\epsilon \xi(x)] = 0$. Similarly, if $\xi_{\max} > \kappa \implies \lim_{\epsilon \to 0} \epsilon \ln[\epsilon \xi] = 0$.

- Secondly, if $\xi(x) < \kappa$ then from (5.11) we obtain

$$\epsilon \ln[_\epsilon \xi(x)] = \epsilon \ln \kappa + \epsilon \ln\{[1 + t^{\frac{1}{\epsilon}}(x)]^\epsilon - 1\}, \qquad (A.21)$$

with $t(x) = \frac{\xi(x)}{\kappa} < 1$. By the Taylor formula of the function $z \to (1+z)^\epsilon$ in the neighborhood of $z = 0$, we have

$$(1+z)^\epsilon = 1 + \epsilon z + \frac{1}{2}\epsilon(\epsilon - 1)z^2 + o(z^2).$$

Hence from (A.21) we obtain

$$\epsilon \ln[_\epsilon \xi(x)] = \epsilon \ln \kappa + \epsilon \ln\{\epsilon t^{\frac{1}{\epsilon}}(x) + \frac{1}{2}\epsilon(\epsilon-1)t^{\frac{2}{\epsilon}}(x) + o(t^{\frac{2}{\epsilon}}(x))\}$$

$$= \epsilon \ln \kappa + \epsilon \ln\{\epsilon t^{\frac{1}{\epsilon}}(x)\} + \epsilon \ln\{1 + \frac{1}{2}(\epsilon-1)t^{\frac{1}{\epsilon}}(x) + o(t^{\frac{1}{\epsilon}}(x))\}$$

$$\longrightarrow_{\epsilon \to 0} \ln t(x) = \ln \frac{\xi(x)}{\kappa}.$$

Similarly, if $\xi_{\max} < \kappa$ then from (5.12) we obtain

$$\lim_{\epsilon \to 0} \epsilon \ln[_\epsilon \xi] = \lim_{\epsilon \to 0} \epsilon \ln\{[1 + \sum_{x \in \mathcal{X}} t^{\frac{1}{\epsilon}}(x)]^\epsilon - 1\}$$

$$= \lim_{\epsilon \to 0} \ln[\sum_{x \in \mathcal{X}} t^{\frac{1}{\epsilon}}(x)]^\epsilon = \ln\left[\frac{\xi_{\max}}{\kappa}\right].$$

Thus finally we have the entries of $V_\kappa(\cdot)$ as: If $\xi(x) > \kappa$, then

$$\lim_{\epsilon \to 0} \ln \left(\frac{_\epsilon \xi}{_\epsilon \xi(x)}\right)^\epsilon = 0.$$

If $\xi(x) < \kappa < \xi_{\max}$, then

$$\lim_{\epsilon \to 0} \ln \left(\frac{_\epsilon \xi}{_\epsilon \xi(x)}\right)^\epsilon = \ln \frac{\kappa}{\xi(x)} > 0.$$

If $\xi_{\max} < \kappa$, then

$$\lim_{\epsilon \to 0} \ln \left(\frac{_\epsilon \xi}{_\epsilon \xi(x)}\right)^\epsilon = \ln \frac{\xi_{\max}}{\xi(x)} > 0.$$

The same case distinction applies for the limit derivatives of $\ln[_\epsilon \xi_\alpha]$ and of $\ln {}_\epsilon \xi_\alpha(u)$ which are required for derivation of the second term in (A.20):

i)
$$\lim_{\epsilon \to 0} \lim_{\alpha \to 0} \frac{\partial}{\partial \alpha} \ln[_\epsilon \xi_\alpha(x)] = \frac{\xi'(x) - \xi(x)}{\xi(x) - \kappa} \quad \text{if } \xi(x) > \kappa,$$

ii)
$$\lim_{\epsilon \to 0} \lim_{\alpha \to 0} \epsilon \frac{\partial}{\partial \alpha} \ln[_\epsilon \xi_\alpha(x)] = \frac{\xi'(x)}{\xi(x)} - 1 \quad \text{if } \xi(x) < \kappa,$$

iii)
$$\lim_{\epsilon \to 0} \lim_{\alpha \to 0} \frac{\partial}{\partial \alpha} \ln[_\epsilon \xi_\alpha] = \frac{\xi_{\max}}{\xi_{\max} - \kappa} \{ E_{B_\xi} \left[\frac{\xi'(\cdot)}{\xi(\cdot)} \right] - 1 \} \quad \text{if } \xi_{\max} > \kappa,$$

iv)
$$\lim_{\epsilon \to 0} \lim_{\alpha \to 0} \epsilon \frac{\partial}{\partial \alpha} \ln[_\epsilon \xi_\alpha] = E_{B_\xi} \left[\frac{\xi'(\cdot)}{\xi(\cdot)} \right] - 1 \quad \text{if } \xi_{\max} < \kappa.$$

The limits above can be derived as follows. According to (5.11) we have
$$\frac{\partial}{\partial \alpha} \ln[_\epsilon \xi_\alpha] = \frac{\partial}{\partial \alpha} \ln\{[h_\epsilon(\xi_\alpha)]^\epsilon - \kappa\},$$

where $h_\epsilon(\xi) = \kappa^{\frac{1}{\epsilon}} + \sum_{x \in \mathcal{X}} \xi(x)^{\frac{1}{\epsilon}}$.

By direct differentiation we obtain
$$\frac{\partial}{\partial \alpha} \ln[_\epsilon \xi_\alpha] = \frac{1}{[h_\epsilon(\xi_\alpha)]^\epsilon - \kappa} [h_\epsilon(\xi_\alpha)]^{\epsilon-1} \sum_{x \in \mathcal{X}} \xi_\alpha^{\frac{1}{\epsilon}-1}(x) \frac{\partial \xi_\alpha(x)}{\partial \alpha}$$

$$\xrightarrow{\alpha \to 0} \frac{[h_\epsilon(\xi)]^\epsilon}{[h_\epsilon(\xi)]^\epsilon - \kappa} \frac{\sum_{x \in \mathcal{X}} \xi^{\frac{1}{\epsilon}}(x)}{h_\epsilon(\xi)} E_\xi^{(\frac{1}{\epsilon})} \left[\frac{\xi'(\cdot)}{\xi(\cdot)} - 1 \right], \quad (A.22)$$

where $E_\xi^{(\frac{1}{\epsilon})}$ denotes the weighted mean with weights equal to $\frac{\xi^{\frac{1}{\epsilon}}(x)}{\sum_{u \in \mathcal{X}} \xi^{\frac{1}{\epsilon}}(u)}$.

Similarly
$$\lim_{\alpha \to 0} \frac{\partial}{\partial \alpha} \ln[_\epsilon \xi_\alpha(x)] = \frac{[l_\epsilon(x)]^\epsilon}{[l_\epsilon(x)]^\epsilon - \kappa} \frac{\xi^{\frac{1}{\epsilon}}(x)}{l_\epsilon(x)} \left[\frac{\xi'(x)}{\xi(x)} - 1 \right], \quad (A.23)$$

where $l_\epsilon(x) = \kappa^{\frac{1}{\epsilon}} + \xi^{\frac{1}{\epsilon}}(x)$.

i) Let $\xi_{\max} > \kappa$. Then we obtain directly from (A.22)
$$\lim_{\epsilon \to 0} \lim_{\alpha \to 0} \frac{\partial}{\partial \alpha} \ln[_\epsilon \xi_\alpha] = \frac{\xi_{\max}}{\xi_{\max} - \kappa} E_{B_\xi} \left[\frac{\xi'(\cdot)}{\xi(\cdot)} - 1 \right].$$

ii) Similarly, if $\xi(x) > \kappa$ we obtain from (A.23)

$$\lim_{\epsilon \to 0} \lim_{\alpha \to 0} \frac{\partial}{\partial \alpha} \ln[_\epsilon \xi_\alpha(x)] = \frac{\xi'(x) - \xi(x)}{\xi(x) - \kappa}.$$

iii) Suppose that $\xi_{\max} < \kappa$, and denote again $t(x) = \frac{\xi(x)}{\kappa} < 1$. From (A.22) we obtain

$$\lim_{\alpha \to 0} \frac{\partial}{\partial \alpha} \epsilon \ln[_\epsilon \xi_\alpha] = \frac{[1 + \sum_{x \in \mathcal{X}} t^{\frac{1}{\epsilon}}(x)]^\epsilon}{[1 + \sum_{x \in \mathcal{X}} t^{\frac{1}{\epsilon}}(x)]^\epsilon - 1} \times$$

$$\frac{\epsilon \sum_{x \in \mathcal{X}} t^{\frac{1}{\epsilon}}(x)}{[1 + \sum_{x \in \mathcal{X}} t^{\frac{1}{\epsilon}}(x)]} E_\xi^{(\frac{1}{\epsilon})} \left[\frac{\xi'(\cdot)}{\xi(\cdot)} - 1 \right]$$

$$\longrightarrow_{\epsilon \to 0} E_{B_\xi} \left[\frac{\xi'(\cdot)}{\xi(\cdot)} - 1 \right] \lim_{\epsilon \to 0} \frac{\epsilon \sum_{x \in \mathcal{X}} t^{\frac{1}{\epsilon}}(x)}{[1 + \sum_{x \in \mathcal{X}} t^{\frac{1}{\epsilon}}(x)]^\epsilon - 1}.$$

Using the Taylor formula, like in the proof above, we obtain that the last limit is equal to 1.

iv) Suppose now, that $\xi(x) < \kappa$. Then from (A.23) we obtain

$$\lim_{\alpha \to 0} \frac{\partial}{\partial \alpha} \epsilon \ln[_\epsilon \xi_\alpha(x)] = \epsilon \frac{[1 + t^{\frac{1}{\epsilon}}(x)]^\epsilon}{[1 + t^{\frac{1}{\epsilon}}(x)]^\epsilon - 1} \times \frac{t^{\frac{1}{\epsilon}}(x)}{1 + t^{\frac{1}{\epsilon}}(x)} \left[\frac{\xi'(x)}{\xi(x)} - 1 \right]$$

$$\longrightarrow_{\epsilon \to 0} \left[\frac{\xi'(x)}{\xi(x)} - 1 \right] \lim_{\epsilon \to 0} \frac{\epsilon t^{\frac{1}{\epsilon}}(x)}{[1 + t^{\frac{1}{\epsilon}}(x)]^\epsilon - 1},$$

and the last limit is equal to 1, which can be proved in the same way as in iii). □

Unfortunately we do not only have a case distinction but also for the cases that $\xi(x) < \kappa$ or $\xi_{\max} < \kappa$, the logarithm in the limits is multiplied by the number ϵ, which in the limit tends to zero. So in this case the limit derivative in (A.20) is infinitely larger than for the case $\xi(x) > \kappa$, $\xi_{\max} > \kappa$. Instead of (A.20) we need to compute

$$\lim_{\epsilon \to 0} \lim_{\alpha \to 0} \frac{1}{\epsilon} \frac{\partial}{\partial \alpha} \Phi[M_{2,\kappa}^{(\epsilon)}(\xi_\alpha)].$$

This makes it impossible to express the derivative as a common sensitivity function and causes the rather complicated structure (5.16) and (5.17) of the algorithm which is described in Section 5.3.

A.5.5 A Simplified Proof for the Condition on Uniformly Optimum Designs (Müller and Pázman (1995))

Take any $A \supseteq A^*$, where A^* corresponds to a uniformly optimum design, then from the condition

$$\dot\eta^T(A) = \sum_{x' \in A^*} c(A,x') f^T(x')$$

it follows that

$$\begin{aligned} M(A) &= \dot\eta(A) C^-(A) \dot\eta^T(A) \\ &= \sum_{x \in A^*} \sum_{x' \in A^*} f(x) c(x,A) C^-(A) c(A,x') f^T(x') \\ &= \sum_{x \in A^*} \sum_{x' \in A^*} f(x) \{C(A)\}_{x.} C^-(A) \{C(A)\}_{.x'} f^T(x') \\ &= \sum_{x \in A^*} \sum_{x' \in A^*} f(x) c(x,x') f^T(x'). \end{aligned}$$

Here, evidently $\dot\eta(A) = \{\dot\eta(x_1), \ldots, \dot\eta(x_{n_A})\}$ and $c(A,x) = \{c(x_1,x), \ldots, c(x_{n_A},x)\} = c^T(x,A)$.

Since this equality holds also for the particular case $A = A^*$, we have $M(A) = M(A^*)$ for any $A \supseteq A^*$. Suppose now that $B \subset \mathcal{X}$ is an arbitrary design. Then, since $A^* \subset B \cup A^*$ the relation

$$\Phi[M(B)] \leq \Phi[M(B \cup A^*)] = \Phi[M(A^*)]$$

must hold for every design B. Hence A^* is optimal for any (concave) criterion $\Phi[\cdot]$. □

A.5.6 Replications in Time

If the replications of the random field are independent in time, we can rewrite the linearized response in stacked form as

$$\dot\eta = \begin{pmatrix} \dot\eta_1(x_{1,1}) \\ \vdots \\ \dot\eta_1(x_{1,n_1}) \\ \dot\eta_2(x_{2,1}) \\ \vdots \\ \dot\eta_T(x_{T,n_T}) \end{pmatrix},$$

with a block-diagonal variance-covariance matrix

$$C(A) = \begin{bmatrix} C(A_1) & 0 & \cdots & 0 \\ 0 & C(A_2) & \cdots & 0 \\ \vdots & & \ddots & \vdots \\ 0 & \cdots & \cdots & C(A_T) \end{bmatrix}.$$

Here $A_t = \{x_{t,1}, \ldots, x_{t,n_t}\}$ and $A = \{A_1, \ldots, A_T\}$, where A_t may be an empty set for some t. By the block inversion rule we have

$$C^{-1}(A) = \begin{bmatrix} C^{-1}(A_1) & 0 & \cdots & 0 \\ 0 & C^{-1}(A_2) & \cdots & 0 \\ \vdots & & \ddots & \vdots \\ 0 & \cdots & \cdots & C^{-1}(A_T) \end{bmatrix},$$

and it follows that

$$M(A) = \sum_{t=1}^{T} \frac{1}{n_A} \sum_{x,x' \in A_t} \dot{\eta}_t(x)[C^{-1}(A_t)]_{x,x'} \dot{\eta}_t^T(x').$$

Assuming $n_A = n_{A_1} = \ldots = n_{A_T}$ we can replace the derivative in (5.6) by a sum of derivatives and consequently use in the algorithm

$$a(x) = \sum_{t=1}^{T} \sum_{z \in \bar{\mathcal{X}}} [M_\kappa(\xi_t^{(s)})]_{x,z} \dot{\eta}(z),$$

$$g(\xi_{(s)}, x) = \sum_{t=1}^{T} a^T(x) \nabla \Phi \left[\sum_{x,z \in \bar{\mathcal{X}}} \dot{\eta}(x)[M_\kappa(\xi_t^{(s)})]_{x,z} \dot{\eta}^T(z) \right] a(x),$$

where ξ_t denotes the design for a particular time point t. □

A.5.7 Equivalence to a Discrete Design Problem
Pázman and Müller (2000)

For the proof of the stated equivalence we require two further properties of $M_2^{(\epsilon)}(\xi)$, namely:

PROPERTY 1. For every $\epsilon > 0$ and every design A we have

$$M_2^{(\epsilon)}(\xi_A) = M(A),$$

where ξ_A is a design measure which is uniform on A and zero outside. To show this property just verify that $\xi_A(x) = (\xi_A)_{\max}$ for every $x \in A$ and use the continuity of the mapping $\xi \to M_2^{(\epsilon)}(\xi)$ (A.5.2).

PROPERTY 2. For every $\gamma > 0$ and every design measure ξ we have

$$M_2^{(\epsilon)}(\xi) \le M(S_\xi)$$

in the Loewner ordering. This follows from the fact that $M(S_\xi) = M_2^{(0)}(\xi)$, and that the additional noise $\epsilon^+(\cdot)$ diminishes the amount of information.

To show that (5.9) is indeed equivalent to (5.2) we let n and $\epsilon \ge 0$ be fixed and suppose that $\max_{A:n_A \le n} \Phi[M(A)] < \infty$.

Then

a) If

$$\xi^\# \in \arg\max_{\xi \in \Xi_n} \Phi[M_2^{(\epsilon)}(\xi)] \qquad (A.24)$$

and $S = S_{\xi^\#}$, then also ξ_S solves (A.24) and

$$S \in \arg\min_{A:n_A \le n} \Phi[M(A)]. \qquad (A.25)$$

b) If $A^\#$ solves (A.25) then $\xi_{A^\#}$ solves (A.24).

If (A.24) holds then for any exact design A with $n_A \le n$, we have

$$\Phi[M(S)] = \Phi[M_2^{(\epsilon)}(\xi_S)] \ge \Phi[M_2^{(\epsilon)}(\xi^\#)] \ge \Phi[M_2^{(\epsilon)}(\xi_A)] = \Phi[M(A)],$$

where we used Properties 1 and 2 — so (A.25) holds, and ξ_S solves (A.24).

Conversely, suppose that $A^\#$ solves (A.25). Then for every $\xi \in \Xi_n$ we have by Properties 1 and 2

$$\Phi[M_2^{(\epsilon)}(\xi)] \le \Phi[M(S_\xi)] \le \Phi[M(A^\#)] = \Phi[M_2^{(\epsilon)}(\xi_{A^\#})],$$

so $\xi_{A^\#}$ is optimal in the sense of (A.24). $\square$

A.6 Proofs for Chapter 6

A.6.1 Simplified Augmentation Rule (Müller and Zimmerman (1997))

From the block matrix inversion rule and the relationship (see e.g. Rao (1973))

$$(A + BCB^T)^{-1} = A^{-1} - A^{-1}B(C^{-1} + B^T C^{-1} B)^{-1} B^T A^{-1}$$

it follows that we can rewrite (6.3) as

$$|M(\xi_{n+1})| = |M(\xi_n) + G(x,\xi_n)V(x,\xi_n)G^T(x,\xi_n)|, \quad (A.26)$$

where

$$G(x,\xi_n) = \dot{\gamma}(\xi_n)\Sigma^{-1}(\xi_n)\Sigma(x,\xi_n) - \dot{\gamma}(x),$$

and

$$V(x,\xi_n) = [\Sigma(x) - \Sigma^T(x,\xi_n)\Sigma^{-1}(\xi_n)\Sigma(x,\xi_n)]^{-1}$$

from the block decomposition

$$\Sigma(\xi_{n+1},\hat{\theta}_0) = \begin{pmatrix} \Sigma(\xi_n) & \Sigma(x,\xi_n) \\ \Sigma^T(x,\xi_n) & \Sigma(x) \end{pmatrix}.$$

From the determinant formula (see e.g. Dhrymes (1984)) for square matrices A and C

$$|A + BCB^T| = |A||C||C^{-1} + B^T A^{-1} B|$$

it is evident that we can rewrite (A.26) as

$$|M(\xi_{n+1})| = |M(\xi_n)||V(x,\xi_n)||V^{-1}(x,\xi_n) + G^T(x,\xi_n)M^{-1}(\xi_n)G(x,\xi_n)|,$$

and for the purpose of maximization the constant term $|M(\xi_n)|$ can be dropped. □

A.6.2. Additional Number of Large Lags when $n - n_0$ Close Points are Augmented (Müller and Zimmerman (1997))

The additional number of large lags can be calculated by subtracting $(n_0 - 1)(n - n_0)$ from the total number of new lags

$$\binom{n}{2} - \binom{n_0}{2} - (n_0 - 1)(n - n_0)$$

$$= \frac{1}{2}[n(n-1) - n_0(n_0-1) - 2(n_0-1)(n-n_0)]$$

$$= \frac{1}{2}[n^2 + n + n_0^2 - n_0 - 2n_0 n]$$

$$= \binom{n - n_0 + 1}{2}.$$

□

174   Appendix

*A.6.3. The Rule of Thumb $n_0 \simeq 0.3n$
(Müller and Zimmerman (1997))*

For a reasonably large n we can approximate

$$n + \frac{1}{2} - \frac{1}{2}\sqrt{2n^2 - 2n + 1} \simeq n + \frac{1}{2} - \frac{1}{\sqrt{2}}\left(n - \frac{1}{2}\right) \simeq n\left(1 - \frac{1}{\sqrt{2}}\right)$$

□.

A.7   D2PT Description

The software used for calculating many of the examples in this monograph is **D2PT**. It was programmed following the general construction principles of so-called application modules in GAUSS (see Aptech (1993)). Therefore it can be directly used under the GAUSS platform. It is based upon an application module for higher dimensional problems in the uncorrelated setup, called MODOPT. A free-of-charge copy of D2PT is available by download from

statistik.wu-wien.ac.at\stat4\mueller\CSD\d2pt.zip

```
/*
D2PT.SRC    Ver. 1.0
(C) Copyright 1997 by Werner G. Mueller,
Department of Statistics,
University of Economics and Business Administration,
                                Vienna, Austria.

Based upon MODOPT Ver. 1.1 by Werner G. Mueller
                assistance by Martin Hemetsberger).

Purpose:   Computes optimal two-dimensional designs.


    FORMAT

    { spectd,weightd,critvd }
        = d2pt(&response,&crit,&algo,&region,points,iterat);
```

INPUT

&response   scalar, pointer to a user provided procedure that represents the functional relationship between the response and the explanatory variables (regression function). How to define &response is illustrated by the following simple example:

    2nd order polynomial without interaction term

```
proc respl(x1,x2,b);
retp(b[1]+b[2]*x1[.]+b[3]*x2[.]
        +t[4]*x1[.]^2+b[5]*x2[.]^2);
endp;
```

    As is evident _mdnth=5 (number of rows of vector b[.]) in our example. (see below)

The user-provided procedure has three input arguments, successively, two coordinate vectors x1[1:n] and x2[1:n], and a vector of parameters b, set by _dth (see below), and a single output argument, a vector of n function values for each of the n points x evaluated at the vector of parameters b.

    A nonlinear example:

```
proc respnl(x1,x2,b);
retp(b[3]*(exp(-b[2]*x1[.])-exp(-b[1]*x2[.])));
endp;
```

    Here evidently _mdnth=3.

For problems nonlinear in the parameters, optimum designs depend on the parameter values, thus for calculation of locally optimum designs prior estimates of the parameters are required, which have to be specified by the vector _dth.

&crit   scalar, pointer to a procedure needed for choosing one of the several possible design criteria. The following criteria are currently at disposal:

         &d_crit....D-criterion,
         &a_crit....A-criterion.

&algo   scalar, pointer to a procedure defining the algorithm used for stepwise generation of near-optimum designs. So far, two algorithms are implemented:

     &goex...computes exact designs (Fedorov,89),
     &goit...computes continuous designs (Wynn,72).

&region   scalar, pointer to a procedure.
The collection of points of the factor space, where measurements are possible and do make sense is called the experimental region. Only points within the chosen region are treated as candidate design points. The following regions are currently supported:

         &cube......unit square
         &sphere....unit circle
         &irreg.....irregular (predefined) region

points   scalar, initial number of observations.
number of initial design points. Replicate points are allowed as long as the number of spectrum points exceeds the number of parameters in the model. For discrete designs, the initial number of observations equals the total number of observations after the optimization process.

A.7 D2PT Description

iterat   number of times the optimization routine defined by &algo will be called. To evaluate the quality of some predefined design, set iterat to 0;

OUTPUT

spectm   N x 2 matrix, containing spectrum points

weightm   Nx1 vector, spectrum point weights. Can take any real number between zero and one. Summation over all weights gives 1.

critm   scalar, (inverse) value of design criterion. Given two designs, the better one has the smaller value. The criterion value corresponding to some optimum design is, of course, problem dependent. So, the absolute optimal criterion value can be quite large.

GLOBAL VARIABLES

_dnth   scalar, number of model parameters, given by the user-provided response function. (see above) No default, must be set by the user.

_dth   _dnth x 1 matrix, prior parameter guesses, Default=0.
If default 0 the prior guess for each parameter in the model is set to 1. Owing to the fact that optimum designs for linear models do not depend on the parameter values, changing the default values is not required. For nonlinear problems, plausible prior guesses should be chosen by the user.

_dgd   2 x 1 matrix, grid density, number of grid points per factor. Suppose _dgd is the 2x1

vector {11,11}. Thus we have 11 grid factor values per factor, giving a total of 11^2 grid points and so optimization is performed by searching over the set of the 121 grid points (assuming a cubical region).
_dgd also determines the distance between neighboring grid points for each of the factors. In our example above, assuming that each of the two factors ranges from -1 to 1, this distance is 0.2 for both factors resulting in the 11 grid values -1.0,-0.8,.....,0.8,1.0 per factor. Default=0. By default _dgd is set to 11 for both factors.

_ded   scalar, minimum distance between spectrum points. Considering the iterative process of design improvement _ded ensures, that any admissible grid point added to the current spectrum has a minimum distance of _ded to any other point so far included in the current spectrum. What is to be avoided by setting _ded are near optimum designs with a large number of points, many of them with small weights and distance to each other, indicating that these are points scattered around the optimal points with much larger weights. What needs to be pointed out is, that setting _ded to another than the default value does only make sense in case of fine grids (what, in most cases, is not a good choice).
Default = -1. If default, D2PT chooses the factor with least distance (LD) between neighbored grid factor values, subtracting LD by 1e-8 to get _ded.

_dxi, _dp

Any design is determined by _dxi in combination with _dp. It can be expressed as the collection of pairs

$$\begin{array}{cc} \_dxi & \_dp \\ x1,x2,\ldots,xn, & p1,p2,\ldots,pn, \end{array}$$

where p1+p2+.....+pn adds up to 1

and pi = ni/N (i = 1,2,...,n)

The n x 2 matrix _dxi contains the x1,x2,...,xn spectrum points and the n x 1 vector _dp contains the p1,p2,...,pn corresponding point weights.

_dxi scalar or nxm matrix containing spectrum of current design. For starting computation some initial design is needed. This can be achieved in two ways:
  1. The initial design is randomly selected,
  2. The initial design is provided by the user.
To randomly select an initial design, set _dxi to zero, which is the default too.

To evaluate the quality of some predefined design, specify _dxi and set iterat (see above) to 0. Default = 0, i.e., random selection of initial design.

_dp scalar 0 or nx1 vector, spectrum point weights, automatically defined if the initial design is randomly selected. If scalar 0, all points will be weighted equally. Default = 0.
For initial designs provided by the user there must be one element in _dp for each of the n spectrum points in _dxi, whereas setting _dp to 0 is, of course, also possible in this case. For discrete designs, the weight of some given spectrum point equals the number of trials at this point divided by the total number of trials

(N).

_dD   _dnth x _dnth matrix, contains current covariance matrix of parameters.

_dA   _dnth x s, matrix of interest.
Default=0. If default,_dA is set to eye(_dnth), the identity matrix, i.e., the user is assumed to be interested equally in all parameters.

_dmax   scalar, pointer to a procedure.
In each iteration, MODOPT tries to find the "best" grid point to add to some current design for design improvement. The method used to fulfill this task is chosen by setting _dmax. Two procedures are available:

    &lines.............. uses a random component
    &direct............. exhaustive search through all admissible grid points.

In almost all cases, the designs found when employing &direct are not better than those found by &lines, but the searching process is much faster when using &lines. So, for reasons of efficiency, &lines will be the better choice in many cases. No default, must be set by the user.

_dregc   scalar 0 or _dgd[1]x_dgd[2] matrix, needed to support procedure &irreg, contains irregular experimental region.

*/

References

Aptech, Inc., (1993). *GAUSS-386 Command Reference*. Washington.

Dhrymes, P.J. (1984). *Mathematics for Econometrics*. Springer Verlag, New York.

Fedorov, V.V. (1974). Regression problems with controllable variables subject to error. *Biometrika*, 61:49–56.

Fedorov, V.V. and Müller, W.G. (1989). Comparison of two approaches in the optimal design of an observation network. *Statistics*, 20 3:339–351.

Fedorov, V.V. and Müller, W.G. (1997). A reparametrization view of optimal design for the extremal point in polynomial regression. *Metrika*, 46:147–157.

Golub, G. and Van Loan, C. (1983). *Matrix computations*. Johns Hopkins University Press, USA.

Hill, P.D.H. (1980). D-optimal design for partially nonlinear models. *Technometrics*, 16:425–434.

Kubáček, L., Kubáčkova, L. and Volaufová, J. (1995). *Statistical Models with Linear Structures*. Veda Publishing House.

Müller, W.G. (1991). On Moving Local Regression. *Unpublished doctoral thesis at the University of Vienna*.

Müller, W.G. (1999). Least squares fitting from the variogram cloud. *Statistics & Probability Letters*, 43:93–98.

Müller, W.G. (2000). Coffee-house designs. In Atkinson, A.C., Bogacka, B., and Zhigljavsky, A.A., editors, *Optimum Design 2000*, forthcoming. Kluwer.

Müller, W.G. and Pázman, A. (1995). Design measures and extended information matrices. Technical Report 47, University of Economics, Department of Statistics, Vienna.

Müller, W.G. and Pázman, A. (1996). Design measures and extended information matrices for optimal designs when the observations are correlated II. Technical Report 48, University of Economics, Department of Statistics, Vienna.

Müller, W.G. and Pázman, A. (1998). Design measures and approximate information matrices for experiments without replications. *Journal of Statistical Planning and Inference*, 71:349–362.

Müller, W.G. and Zimmerman, D.L. (1995). An algorithm for sampling optimization for semivariogram estimation. In Kitsos, C.P. and Müller, W.G., editors, *Model-Oriented Data Analysis 4*, Heidelberg. Physica, 173–178.

Müller, W.G. and Zimmerman, D.L. (1997). Optimal design for semivariogram estimation. Technical report #51, University of Economics, Department of Statistics, Vienna.

Pázman, A. (1993). *Nonlinear Statistical Models*. Mathematics and Its Applications. Kluwer Academic Publishers, Dordrecht.

Pázman, A. and Müller, W.G. (1998). A new interpretation of design measures. In *Model-Oriented Data Analysis 5*, Atkinson, A.C., Pronzato, L., and Wynn, H.P., editors. Physica-Verlag, Heidelberg.

Pázman, A. and Müller, W.G. (2000). Optimal Design of Experiments Subject to Correlated Errors, *Statistics & Probability Letters*, forthcoming.

Pukelsheim, F. (1993). *Optimal Design of Experiments*. John Wiley & Sons, Inc., New York.

Rao, C.R. (1973). *Linear Statistical Inference and Its Applications*. Wiley, New York, 2nd edition.

Ripley, B.D. (1981). *Spatial Statistics*. Wiley, New York.

List of Figures

1.1 The Upper-Austrian SO_2 monitoring network; circles represent sites. 6
1.2 The water quality monitoring network in the Südliche Tullnerfeld; solid circles represent sites, grid points represent region. 6

2.1 A generic tricube (solid) and McLain's (dashed) weight function. 18
2.2 A generic spherical variogram (h horizontal, γ vertical). 22
2.3 Daily averages of SO_2 concentrations in mg/m^3 in Steyregg-Weih (horizontal time, vertical SO_2). 26
2.4 Contour plot of a local regression estimate of the trend surface. 27
2.5 Variogram cloud (dots) and various variogram estimators for the days 94-02-09, 94-03-01, 94-03-16, 94-12-16 (from upper right to lower left, vertical γ, horizontal h). 28
2.6 Squared simple kriging contour plot of the detrended data. 29

3.1 Confidence ellipsoids corresponding to different designs . 45
3.2 Induced design space $\mathcal{F}$ and intersection points. 47

3.3	Contour map of the sensitivity function on a unit square.	48
3.4	Confidence ellipsoids for various designs.	49
3.5	Sensitivity function (vertical) for the extremal value problem on the unit square.	55
3.6	A locally D-optimum design.	59
3.7	A D_L-optimum design for localizing the peak.	60
3.8	A D-optimum design for estimating the mean parameter in a random coefficient model - a Bayesian D-optimum design	61
4.1	A 21 point Fibonacci lattice on the unit square.	73
4.2	Two 6-point maximin distance designs on the unit circle (small circles and small solid circles represent sites).	75
4.3	A 17-point 'coffee-house' design.	82
4.4	A design for local regression with a tricube weight function and $\hat{\kappa} = 0.7$.	84
4.5	A design for discriminating a second order polynomial from a Gaussian p.d.f.	85
5.1	A replication-free optimum design for linear regression	98
5.2	D-optimum design for a correlated example.	105
5.3	A locally D-optimum replication-free design.	116
5.4	A locally D-optimum design under a given correlation structure.	117
5.5	A Φ-optimum design under a given correlation structure with replications in time (the smaller circle indicates an unreplicated point).	118
6.1	A D-optimum design for linear regression under heteroscedasticity.	127
6.2	A generic D-optimum design for a spherical variogram in the lag space (bars stand for design weights p_i and the curve represents the sensitivity function).	135
6.3	A D-'optimum' design for variogram estimation.	142
6.4	A design for variogram estimation when the lag correlations are ignored.	143
6.5	Compound criterion values $\bar{\Phi}[\cdot]$ vs. n_0 for $\alpha = 0.5$ (solid line), $\alpha = 0.25$ (dashed line) and $\alpha = 0.75$ (closely dashed line).	144
6.6	A 'compound' optimum design (with $n_0 = 8$).	145

Author Index

Abt, M., 110, 121, 130, 147
Altman, N.S., 21, 32
Angulo, J.M., 147
Arbia, G., 2, 8, 113, 121
Atkinson, A.C., 2, 8, 38, 43, 44,
    46, 53, 63, 66, 80, 81,
    87, 89, 121, 123, 127,
    147, 181, 182
Atwood, C.L., 51, 63

Baafi, E., 35
Bandemer, H., 38, 63
Barnett, V., 89
Bates, R.A., 72, 87
Batsell, S., 114, 121
Behar, J.V., 90, 124
Bellhouse, D.R., 2, 8, 72, 87
Bellmann, A., 63, 110, 121
Benedetti, R., 113, 121
Bickel, P.J., 111, 121
Bischoff, W., 109, 121
Boer, E.P.J., 123

Bogacka, B., 66, 89, 181
Bogaert, P., 132, 147
Bogardi, W.I., 123
Boltze, L., 110, 121
Borth, D.M., 82, 87
Box, G.E.P., 25, 32, 37, 38, 43,
    45, 63, 69–72, 87
Boyd, H.A., 35
Brimkulov, U.N., 112, 121, 132,
    147
Brockwell, P., 10, 32
Buck, R.J., 87
Bueso, M.C., 139, 147
Buja, A., 20, 32
Buonaccorsi, J.P., 54, 63
Burgess, T.M., 122, 149

Cambanis, S., 73, 87, 111, 123
Caselton, W.F., 139, 147
Cassel, C., 11, 32
Chaloner, K., 42, 56, 63, 64
Chambers, J.M., 32

Chang, Y.-J., 70, 87
Cheng, C., 50, 64
Cheng, M., 78, 87
Chernoff, H., 41, 42, 64
Chiles, J., 9, 32
Christakos, G., 2, 8
Cleveland, W.S., 15, 17, 19, 20, 32, 79, 87
Conan-Doyle, A., 91, 121
Cook, R.D., 42, 53, 64, 67, 95, 121, 125–127, 147
Cox, D.D., 1, 8
Cox, D.R., 25, 32, 37, 64
Cox, L.H., 8
Cox, M.A.A., 134, 147
Cox, T.F., 134, 147
Cressie, N., 2, 8, 9, 11, 13–15, 20, 23, 24, 32, 36, 113, 121, 147
Cruz-Sanjulian, J., 147
Currin, C., 139, 147
Curtis, A., 50, 64

DasGupta, A., 56, 64
Davis, R.A., 10, 32
Delfiner, P., 9, 32
Dempster, A.P., 68
Der Megréditchian, G., 57, 64
Dette, H., 49, 50, 64, 125, 148
Dhrymes, P.J., 156, 158, 173, 181
von Doderer, H., 69, 88
Dodge, Y., 148
Donev, A.N., 38, 46, 53, 63, 81, 87
Draper, N.R., 38, 63, 69–71, 87
Droge, B., 18, 32
Dutter, R., 9, 32

Eccleston, J.A., 107, 123

Edwards, L.L., 67
Elfving, G., 47, 64
Elkins, T.A., 35
Ellis, J.H., 140, 150
Ensore, K.B., 8
Eubank, R.L., 20, 33
Evans, J.W., 70, 88

Fan, J., 20, 33
Fang, K.-T., 74, 88
Fedorov, V.V., 2, 8, 12, 13, 15, 16, 23, 24, 33, 38, 40–42, 44, 46, 50, 51, 54–56, 59, 64, 65, 67, 72, 78, 80, 81, 84, 87, 88, 91, 95, 99, 108, 111–114, 121, 122, 125, 126, 132, 140, 147–149, 153, 154, 156, 181
Felsenstein, K., 81, 88
Fienberg, S.E., 2, 8, 33
Firth, D., 42, 65
Fisher, R.A., 37, 65
Flanagan, D., 111, 121, 122, 132, 148
Flatman, G.T., 90, 124
Ford, I., 42, 43, 47, 65
Francis, B.J., 34

Gaffke, N., 52, 53, 65
Garcia-Arostegui, J.L., 147
Genton, M.G., 24, 33
Gladitz, J., 56, 65
Glatzer, E., 112, 122
Goldberger, A., 13, 33
Golub, G., 163, 181
Goodwin, G.C., 67
Gosh, S., 63, 64, 66, 87, 88, 122, 148
Gotway, C.A., 23, 33, 121

Gribik, P., 57, 66
Grononda, M.O., 121
Grosse, E.H., 32
Guttorp, P., 10, 21, 29, 33, 35, 140, 148

Haas, T.C., 20, 28, 33, 138, 148
Hackl, P., 33, 38, 41, 44, 46, 64, 65, 72, 88, 111, 122
Haines, L.M., 43, 46, 49, 53, 63, 64, 66
Haining, R., 1, 8, 11, 33
Hall, P., 34, 87
Hamilton, D.C., 44, 66
Hannan, E.J., 87
Härdle, W., 18, 20, 32, 34
Hastie, T.J., 15, 20, 32, 34
Hatzinger, R., 34
Hedayat, A., 50, 66
Heemink, A.W., 35
Heiligers, B., 52, 64, 65
Hendrix, E.M.T., 123
Herzberg, A.M., 63, 72, 87, 88, 111, 121
Hill, P.D.H., 155, 181
Hinde, J., 42, 65
Holst, U., 35
Homer, K., 130, 133, 150
Host, G., 21, 34
Hössjer, O., 35
Huda, S., 72, 88
Huijbregts, C.J., 22, 34
Hunter, J.S., 63
Hunter, W.G., 63

Isaaks, E.H., 1, 8, 9, 34
Iyer, H.K., 54, 63

Jacobson, E., 23, 34, 128, 148
John, P.W.M., 75, 88

Johnson, M.E., 74, 88
Johnson, N.L., 8
Journel, A.G., 22, 34
Jung, W., 63
Juritz, J., 63

Kelly, W.E., 123
Khabarov, V., 81, 88
Kiefer, J., 38, 39, 50, 51, 66, 92, 122
Kitsos, C.P., 65, 149, 181
Ko, C.W., 139, 148
Koehler, J.R., 72, 88, 139, 148
Kortanek, K., 66
Kotz, S., 8
Krishnaiah, P.R., 87
Krug, G.K., 121, 147
Kubáček, L., 162, 181
Kubáčkova, L., 181

Lafratta, G., 113, 121
Laird, N.M., 68
Lall, U., 18, 35
Lamorey, G., 23, 34, 128, 148
Läuter, E., 125, 148
Läuter, H., 149
Le Anh Son, 63
Le, N.D., 148
Lee, J., 148
Lee, S.Y., 135, 148
Li, K., 50, 64
Loader, C., 15, 32
Lucas, H.L., 45, 63

Maderbacher, M., 16, 34
Maljutov, M.B., 126, 149
Manson, A.R., 70, 88
Marron, J.S., 34
Martin, R.J., 37, 66
Matern, B., 72, 88

Mathar, R., 53, 65
Matheron, G., 1, 8, 21, 34
May, K., 1, 8
McArthur, R.D., 73, 88
McBratney, A.B., 23, 24, 34, 73, 89, 113, 122, 128, 142, 149
McLain, D.H., 17, 34
McRae, J.E., 79, 87
Micchelli, C.A., 113, 122
Miller, A.J., 52, 67
Mirnazari, M.T., 50, 66
Mitchell, T.J., 52, 66, 75, 77, 89, 123, 147, 149
Mizrachi, D., 150
Molchanov, I., 52, 66
Montepiedra, G., 88
Montgomery, D.C., 38, 67
Moore, L.M., 88
Morris, M.D., 75, 77, 89, 129, 147, 149
Müller, H.G., 34, 79, 89
Müller, W.G., 16, 23, 24, 33, 34, 42, 54–56, 64, 65, 67, 77–79, 81, 83, 89, 93, 95, 97, 101, 107, 108, 112, 122, 123, 128, 133–135, 137, 140, 148, 149, 153, 154, 156, 157, 159, 161, 163, 165, 170–174, 181, 182
Müller-Gronbach, T., 110, 122
Munn, R.E., 5, 8, 34
Myers, D.E., 128, 134, 137, 150
Myers, E.W., 135
Myers, R.H., 38, 67

Nachtsheim, C.J., 53, 64, 88, 125, 147
Nagel, S., 63

Nagel, W., 63
Näther, W., 4, 8, 63, 91, 101, 108–110, 112, 114, 121, 122
Ng, T.S., 67
Nguyen, N.K., 52, 67
Notz, W.I., 70, 87
Nychka, D., 75, 76, 89, 90

O'Hagan, A., 80, 89
Oehlert, G.W., 131, 149
Opsomer, J.D., 21, 35
Owen, A.B., 72, 88, 139, 148

Palma, D., 121
Park, S.H., 80, 89
Patil, G.P., 33, 148
Pázman, A., 11, 35, 38, 40, 41, 67, 89, 93, 95, 97, 101, 108, 122, 123, 153, 159, 161, 163, 165, 170, 171, 181, 182
Pedder, M.A., 26, 35
Pelto, C.R., 15, 35
Pesti, G., 113, 123
Pettitt, A.N., 73, 89, 128, 149
Pilz, J., 14, 35, 56, 63, 65, 67
del Pino, G., 12, 32
Pirsig, R., 37
Pitovranov, S.E., 58, 67
Ponce de Leon, A.C., 81, 89
Pötscher, B.M., 42, 67
Prince, 125, 149
Pronzato, L., 41, 67, 121, 123, 182
Pukelsheim, F., 3, 38–40, 53, 67, 82, 89, 95, 100, 123, 159, 182

Quereshi, Z.H., 43, 67

Queyranne, M., 148

Rabinowitz, N., 113, 123, 150
Rafajłowicz, E., 57, 67
Rajagopalan, B., 18, 35
Rao, C.R., 33, 63, 64, 66, 87, 88, 122, 148, 159, 172, 182
Rao, M.M., 87
Rasch, D.A.M.K., 99, 100, 123
Read, C.B., 8
Riccomagno, E., 74, 87, 89
Richter, K., 63
Rieder, S., 53, 67
Ripley, B.D., 9, 15, 35, 153, 182
Roddenberry, G., 9, 35
Rosenberger, J.L., 82, 89
Royle, J.A., 75, 89, 90
Rubin, D.B., 68
Ruppert, D., 24, 35
Russo, D., 128, 132, 147, 149

Sacks, J., 108–110, 123, 129, 132, 139, 149
Sampson, P.D., 10, 21, 29, 33, 35, 148
Särndal, C., 32
Saunders, I.W., 107, 123
Savanov, V.L., 121, 147
Schilling, M.F., 107, 123
Schimek, M.G., 32, 35
Schoefield, N., 35
Schumacher, P., 140, 149
Schwabe, R., 89
Sebastiani, P., 140, 149
Seber, G.A.F., 24, 35
Seeber, G.U.H., 34
Shah, K.R., 50, 68
Shewry, M.C., 139, 149
Shimshoni, Y., 150
Shyu, M.J., 32

Silvey, S.D., 38, 40, 42, 46, 50, 52, 65, 68
Sinha, B.K., 50, 68
Spoeck, G., 35
Spokoiny, V.G., 107, 123
Srivastava, R.M., 1, 8, 9, 34
Steckel-Berger, G., 34
Stein, M.L., 13, 35, 128, 150
Steinberg, D.M., 113, 123, 126, 150
Stone, M., 18, 35
Studden, W.J., 64
Su, Y., 111, 123
Sweigart, J., 66

Tanur, J.M., 2, 8, 33
Thiebaux, H.J., 26, 35
Tibshirani, R.J., 15, 32, 34
Titterington, D.M., 65, 68, 87
Tobias, R., 75, 90
Torsney, B., 52, 65, 68
Trujillo-Ventura, A., 140, 150
Turkman, K.F., 89

Uciński, D., 43, 57, 68

van Eijkeren, J.C.H., 35
Van Loan, C., 163, 181
Verdinelli, I., 42, 56, 63
Volaufová, J., 89, 181
Vuchkov, I., 67

Wackernagel, H., 9, 35
Wahba, G., 80, 90, 110, 113, 122, 124
Walden, A.T., 148
Wand, M.P., 35
Wang, Y., 74, 88
Warrick, A.W., 128, 134, 135, 137, 150

Watts, D.G., 44, 66
Webster, R., 23, 24, 34, 122, 149
Welch, W.J., 50, 66, 123, 130, 147, 149
White, L.V., 50, 68
Whittle, P., 50, 68
Wiens, D.P., 43, 68
Wild, C.J., 24, 35
Wilhelm, A., 53, 68
Wilson, K.B., 37, 63, 72, 87
Wolfowitz, J., 50, 51, 66
Wong, W.K., 126, 147
Wretman, J.H., 32
Wu, C.F.J., 65
Wu, S., 140, 150
Wynn, H.P., 51, 68, 87, 89, 95, 99, 121, 123, 124, 139, 140, 148, 149, 182

Yang, Q., 89
Yfantis, E.A., 72, 90, 113, 124
Ying, Z., 138, 139, 150
Ylvisaker, D., 73, 88, 90, 108–111, 123, 124, 132, 147, 149
Yuditsky, M.I., 107, 124

Zhang, X.F., 23, 35
Zhigljavsky, A.A., 66, 89, 181
Zidek, J.V., 139, 140, 147–150
Zimmerman, D.L., 14, 24, 36, 128, 130, 133–135, 137, 149, 150, 153, 172–174, 181, 182
Zimmerman, M.B., 24, 36
Zuyev, S., 52, 66

Subject Index

additivity, 40, 92, 118
algorithm, 52, 53, 82, 95, 96, 98–100, 102, 103, 105, 112, 113, 115, 116, 128, 132, 133, 137, 141, 155, 164, 169, 171
    'coffee-house', 76
    approximate, 99
    augmentation, 132
    branch and bound, 99
    computational, 38
    constrained design, 42
    convergence of, 100
    correction, 52, 112
    design, 4, 53, 70, 92, 93, 114, 117
    distance, 135–137, 153
    enumeration, 100
    exact, 139
    exact design, 53
    exchange, 4, 53, 99, 100, 132
    Gauss-Newton, 11
    gradient, 60, 95, 96, 103, 108, 120
    greedy, 131
    heuristic, 112
    iterative, 4, 24, 97
    numerical, 51, 83, 95
    o.d.e., 38
    one-point correction, 51, 97, 98, 103, 132
    optimization, 53, 91, 92
    sampling, 23
    search, 107
    sequential, 81
    simulated annealing, 75
    simultaneous correction, 52
alphabetic optimality, 69
anisotropy, 10, 26, 71

best linear unbiased estimator, 11, 91, 110

concentration, 6, 26, 57, 83, 151, 152

confidence ellipsoid, 44–46, 48, 49
constrained measure, 99
convergence, 51, 73, 97, 100, 112, 133
convexity, 40, 92
correlation, 72, 132
    ignoring, 132, 133, 137, 141, 143
    local, 4, 91
    spatial, 10
correlation coefficient, 58
correlation structure, 101, 107, 109, 117, 118
covariance, 111, 114
    function, 14, 40, 73, 105, 108, 110, 115, 117, 126, 132, 138, 139
    function, decreasing, 75
    function, design independent, 109
    function, misspecifying the, 13
    function, parameterized, 10, 14, 140
    function, spatio-temporal, 117
    kernel, 111
    matrix, 19, 23, 57, 94, 109, 131, 132, 156, 157, 171
    structure, 24, 120
criterion, 3, 44, 46, 49, 50, 96, 98, 99, 110, 113, 118, 125, 126, 128, 129, 140–142, 165
    Φ_p-optimality, 50
    α-optimality, 49
    A-optimality, 50, 78
    asymptotic Φ-optimality, 110
    Bayes-A-Optimality, 56
    bias minimization, 70
    compound, 113, 140–142, 144
    concave, 94, 95, 170
    D-optimality, 50, 78, 109, 110, 112, 130, 140
    E-optimality, 49, 50
    entropy, 50
    G-optimality, 49
    increase in, 51
    independence, 136
    J-optimality, 70
    L-optimality, 78
    local, 81
    Loewner, 44
    maximin, 140
    maximum variance optimality, 49
    optimality, 48, 50, 56, 79, 96, 163
    optimization, 79
    T-optimality, 80, 81
    U-optimality, 74
criterion function, 79, 95, 97, 104
cross validation, 18, 26

data
    geostatistical, 14, 24, 127
    spatial, i, viii, 1, 4, 9, 11, 54, 91
    spatio-temporal, 56
data collection, 1, 70
data generating process, 1, 2
data sparsity, 128
data transformation, 24
derivative
    directional, 44, 52, 96, 97, 103, 104, 108, 163, 165
    directional, linear, 97
    finite second, 20
    Gateaux, 52

limit, 167, 169
design
    'coffee-house', 76, 82, 156
    approximate, 40, 46, 48, 52, 53, 81, 157, 162
    balanced incomplete block, 37
    Bayesian, 56, 156
    categorical, 37
    central composite, 3, 38, 72
    compound, 126, 128, 143
    constrained, 95, 113, 126
    continuous, 39
    exact, 39, 48, 52, 53, 56, 95, 99–101, 157, 161
    exploratory, 3, 69–71, 73
    initial, 52, 71
    lattice, 74
    maximin distance, 3, 74–77, 82
    model discriminating, 80, 81
    multipurpose, 4, 125
    product, 73, 111
    replication-free, 92, 95, 98–100, 108, 110, 115, 116, 139
    response surface, 38
    robust, 50, 70
    sequential, 42
    singular, 45
    space-filling, 79, 82, 83, 116
    spatial, 2, 3, 5, 7, 38, 107, 113, 117
design criterion, 3, 44, 47, 53, 54, 69, 70, 80, 82, 86, 93, 110, 112, 117, 118, 125, 126, 129, 134, 137, 139, 141, 142, 144, 155

design measure, 4, 44, 51, 52, 58, 62, 80, 83, 91–95, 100, 101, 120, 127, 160, 162
design region, 38, 46, 53, 55, 58, 72, 74–76, 83, 98
design space, 9, 46, 47, 58, 98, 153
detrending, 14, 24, 25, 114
differential equations
    partial, 43
discrepancy, 74
distance, 3, 16, 17, 21, 46, 76, 82, 100, 116, 128, 133, 134, 136, 154
    average, 17
    Euclidean, 139
    intersite, 4, 10, 15
distributed parameter system, 43, 57

eigenvalue, 50, 111
Elfving set, 47
entropy, 50, 82, 139, 140
equivalence theorem, 50, 51, 56, 80, 95, 99, 100, 112, 114
equivalent degrees of freedom, 20
equivalent number of parameters, 20
extremal
    observation, 54
    point, 54, 55
    value, 3, 54, 55, 154

Fibonacci lattice, 72, 73

generalized least squares estimation, 2, 11, 13, 23, 30

information, 11, 15, 20, 23, 39–42, 56, 73, 81, 92–94, 101, 102, 108, 110, 130, 131, 135, 137, 139
   measure of, 39, 41, 78, 139
   prior, 56, 60, 140
information matrix, 4, 39–42, 44, 62, 70, 71, 91–97, 100, 101, 103, 117, 125, 127, 129, 130, 134, 155, 160, 161, 165
   approximate, 41, 92, 94–96, 100–102, 104, 114, 120, 132, 161, 163, 165
information transfer, 101
isotropy, 10, 16, 128

kriging, viii, 2, 13, 15, 27, 113, 128, 129, 138
   minimax, 15
   nonparametric, 20, 28
   simple, 28, 29
   universal, 2, 12–15, 20, 21, 30, 113, 126
kriging variance, 13, 15, 113, 114

lack of fit, 71, 80, 81
Lagrange method, 96, 101
local optimality, 41, 58, 59, 115–117

model
   assumptions, 62
   building, 3, 70
   discrimination, 71, 83, 125
   misspecification, 70
   validation, 3, 141
monitoring, 1, 4, 5, 25, 31, 83, 116

network, viii, 5, 6, 24, 57, 115, 138, 140, 141
moving sensors, 57

noncentrality parameter, 80
nonconcavity, 95, 98, 100
nonconvexity, 95, 161
nonlinearity, 43, 45, 50, 83, 97, 126, 129, 134, 153
nugget-effect, 21, 22
numerical
   construction of optimum designs, 99
   errors, 103, 107
   minimization, 113
   optimization procedure, 83
   procedure, 81
   stability, 17

optimization, 4, 7, 41, 54, 95, 100, 103, 114, 116, 125, 130, 131, 137, 142

parameter
   estimation, 38, 40, 49, 54, 58, 61, 71, 74, 80, 82–84, 86, 114, 125, 139, 140
   space, 10, 22
prediction, 1, 10, 12, 13, 15, 19, 21, 30, 49, 72, 83, 113, 114, 128, 139
   best linear, 13
   spatial, 30, 113, 128
prediction variance, 3, 112–114

random coefficient model, 3, 55, 57, 59, 61, 111, 114, 140
random field, i, 2, 9, 10, 14, 18, 21, 30, 40, 54, 74, 83,

91, 100, 113, 116, 117, 126, 139, 141, 149, 170
  Gaussian, 23
randomization, 72
range, 22, 23, 27, 115, 116, 134, 137
regression, 20, 37, 91, 92, 109, 110, 129, 132
  Bayesian, 55, 140
  Fourier, 74
  kernel, 79
  least squares, 21
  linear, 42, 98, 105, 107–109, 126, 127, 140
  local, 2, 3, 15, 16, 20, 26, 27, 30, 77–80, 83, 84, 153, 157
  nonparametric, 15, 77, 83
  quadratic, 54
regression function, 3, 20, 30, 108
regression model, 2, 80
replications, 98, 99, 111, 116, 118, 161, 170
  in time, 116, 118, 170
  instantaneous, 3, 92
  number of, 39, 94
response, 3, 16, 45, 49, 54, 69–71, 74, 78, 81, 82, 86
  average, 48
  linear, 13, 46, 47, 80, 170
  local, 16, 19, 79
  local linear, 83
response surface, 43

sampling, 2, 3, 11, 23, 107, 128, 133, 153
  maximum entropy, 139, 140
  random, 3, 50, 71–73, 76
  regular, 73

sensitivity function, 44, 48, 51, 53, 55, 56, 60, 79, 80, 83, 96, 97, 106, 110–112, 127, 135, 140, 157, 169
sill, 22, 27
simulation experiment, 16, 78, 108, 112, 129, 139
smoother, 20, 21, 79
  local linear, 20
smoothing
  kernel, 20
  local polynomial, 20
  methods, 20
  spline, 20
smoothing parameter, 17–20
smoothness, 19, 73, 77
  amount of, 17, 18, 20
smoothness evaluation, 20
spatial
  covariance, 10
  covariance function, 2
  dependence, 14, 21
  observation, 91
  process, 4, 10
  statistics, 2, 9, 15, 40, 70
  trend, 3, 11–13, 25, 77, 91, 126
spherical variogram, 22, 115, 134, 135
support, 46, 48, 51, 52, 55, 74, 93, 95–97, 99, 103, 107, 108, 127, 155, 162
  one-point, 44, 164
  set of, 39, 40
support points, 40, 46, 48, 52, 58, 69, 96, 98, 101, 103, 110, 113, 115

time series, 10, 107

transformation
    linear, 10
    square-root, 25

uniform weights, 17

variogram, 2, 4, 14, 21–23, 71, 113, 120, 126–129, 133, 137
    cloud, 22, 23, 27, 28, 133
    empirical, 128
    estimate, 27, 129
    estimation, viii, 2, 23, 30, 62, 73, 127–130, 137, 138, 141–143
    estimator, 21, 28, 153
    fitting, 4, 12, 15, 21, 27, 125, 129, 137
    model, 15, 21, 22, 127–129, 134
    parameter, 28
    range, 133

weight function, 16, 17, 49, 78, 79, 83, 153
    tricube, 17–20, 26, 83, 84
weighted least squares, 2, 12, 15, 23, 24, 78, 133, 153